ADDITIONAL
APPLIED
MATHEMATICS

L. HARWOOD CLARKE

and

F. G. J. NORTON

Head of the Mathematics Department, Rugby School

SECOND EDITION

HEINEMANN EDUCATIONAL BOOKS

Heinemann Educational Books Ltd
Halley Court, Jordan Hill, Oxford OX2 8EJ

OXFORD LONDON EDINBURGH
MELBOURNE SYDNEY AUCKLAND
IBADAN NAIROBI GABORONE
KINGSTON PORTSMOUTH NH (USA)
SINGAPORE MADRID

9.50

Printed and bound in Great Britain by
Richard Clay Ltd, Bungay, Suffolk

PREFACE

THIS BOOK has been written as a companion volume to *Additional Pure Mathematics*, and covers the Mechanics syllabus of the G.C.E. Alternative Ordinary papers, of Additional Mathematics, and of the Subsidiary Mathematics of the Overseas Higher School Certificate. The various syllabuses differ considerably, and this book contains more than is necessary for any one of these Examinations. Perhaps even more important, though, we hope that this book will be suitable as an introduction to Theoretical Mechanics, whether started in the fifth form or in the first year of Post Certificate work. A second volume, *Advanced Level Applied Mathematics*, covers the rest of the topics generally taught in Applied Mathematics in schools.

One of the most valuable changes made in school syllabuses in recent years has been the widespread use of vectors, and pupils are now being introduced to them at this level. Examining Boards restrict vectors at Additional Mathematics to addition, subtraction and scalar multiples, and this is very reasonable. We have therefore been able to develop velocity, acceleration, force and momentum as vectors, writing the momentum equation

$$t\mathbf{F} = m\mathbf{v} - m\mathbf{u}$$

but we have been compelled to treat the work done by a force without vectors, stopping short of **F**.s. This comes easily in the work of the following year. All this has proved well within the ability of students at this stage, and these chapters have been tested thoroughly in class: some of the later geometrical questions, though, are quite demanding. There is a revision exercise restricted to questions on vectors, some reprinted from recent papers set by G.C.E. boards.

We are greatly indebted to the Second Report on the Teaching of Mechanics by the Mathematical Association, and we share the desire expressed in that report that pupils should be encouraged to look around themselves for applications of the ideas discussed here. Some of these are indicated throughout the book.

We should like to express our most sincere thanks to Professor A. Geary, formerly of The City University, London, for reading the manuscript and for many helpful suggestions: to Mr J. W. Norton, of The King's School, Bruton, for checking the answers and for his kindly fraternal criticism: finally to Mr Hamish Mac-Gibbon, of Heinemanns, for advice and encouragement in the writing of this book.

Questions are reprinted from papers set by several Examinir.g Boards, and we should like to thank these Boards for permission to reprint the questions. Such questions are indicated in the text by the following abbreviations:

Associated Examining Board	A.E.B.
Joint Matriculation Board	J.M.B.
Oxford & Cambridge Examining Board	O. & C.
Oxford & Cambridge Board, M.E.I. project	M.E.I.
Oxford & Cambridge Board, S.M.P. project	S.M.P.
Oxford Delegacy for Local Examinations	O.
Cambridge University Local Examinations Syndicate	C.

Where such questions have been changed into SI units, this is indicated by*.

Rugby, 1972   F. G. J. Norton

PREFACE TO THE SECOND EDITION

The increasing use of vectors since this book was first written and their extension to three dimensions justifies the inclusion of an extra chapter on vectors; the opportunity has also been taken to add another Miscellaneous Exercise, consisting mainly of questions from past G.C.E. papers. As before, I am grateful to the G.C.E. Boards listed above for permission to reprint past questions.

Rugby, 1979   F.G.J.N.

Abbreviations used

s	seconds
mm	millimetres
cm	centimetres
m	metres
km	kilometres
$cm\ s^{-1}$	centimetres per second
$m\ s^{-1}$	metres per second
$km\ h^{-1}$	kilometres per hour
$m\ s^{-2}$	metres per second per second

The SI unit of force is   1 newton (1 N)
       of work or energy is   1 joule (1 J)
       of power is   1 watt (1 W)

1 newton $= 1\ kg\ m\ s^{-2}$
1 joule $= 1$ newton metre, 1 N m
1 watt $= 1$ joule per second, $1\ J\ s^{-1}$

Notation used

Force   Acceleration

Velocity   Resultant velocity
                               (Chapter 13)

CONTENTS

CHAPTER 1

Velocity

Average speed

If a car travels 120 km in 3 hours, we say that its average speed is 40 km h^{-1}. We certainly do not mean that the car is travelling at a constant speed of 40 km h^{-1}; it will doubtless have been stopped by traffic lights unless it is on a motorway, and on clear stretches of road, its speed will have greatly exceeded 40 km h^{-1}. The average speed is found by dividing the total distance travelled by the time taken to cover that distance. Thus I may run 1 km to the railway station at 10 km h^{-1}, and then travel by train the 120 km to London at 120 km h^{-1}. I travel $(1 + 120)$ kilometres in $(\frac{1}{10} + 1)$ hours, so my average speed for the journey is

$$\frac{1 + 120}{\frac{1}{10} + 1} \text{ kilometres per hour,}$$

i.e. $\qquad\qquad$ 110 km h^{-1}.

Notice that this is not the average of 10 km h^{-1} and 120 km h^{-1}.

$$\text{Average speed} = \frac{\text{total distance travelled}}{\text{total time taken}}.$$

Velocity, average velocity

When the *direction* in which we are travelling is taken into account, as well as the speed, we call this our velocity. The car may have a velocity of 60 km h^{-1} in a north-east direction; my velocity to the station is 10 km h^{-1} due north; the train's velocity is 120 km h^{-1} south-east. Velocity tells us how distance is changing with time and is defined

velocity = rate of change of distance measured from
a fixed point.

The average velocity of a body is thus the change in its distance

from a fixed point called the displacement divided by the interval
of time that has elapsed. Concorde may fly from London to
Edinburgh and back at an average speed of 2000 km h^{-1}, but the
average velocity for the round trip will be zero!

Uniform velocity

A body is said to be moving with uniform velocity if it travels
equal distances in a constant direction in equal intervals of time.
If a graph relating displacement and time is drawn, the result
will be a straight line.

Angular velocity

For a rotating body, we generally express the change of position
in terms of angular velocity, which is the size of the angle turned
through in a unit of time. This angle may be measured in revolutions or radians. Thus a shaft may have an angular velocity of
4000 revolutions per minute. The hour hand of a clock turns
through one revolution in 12 hours and so its angular velocity is
$\frac{1}{12}$ of a revolution per hour, or $\pi/6$ radians per hour or $\pi/21\,600$
radians per second.

Example 1. *A train travels 100 km at 50 km h^{-1} and a further
40 km at 60 km h^{-1}. What is the average speed over the 140 km?*

$$100 \text{ km at } 50 \text{ km h}^{-1} \text{ takes } 2 \text{ hours,}$$
$$40 \text{ km at } 60 \text{ km h}^{-1} \text{ takes } \tfrac{2}{3} \text{ hours,}$$

so
$$140 \text{ km is travelled in } 2\tfrac{2}{3} \text{ hours.}$$

$$\therefore \quad \text{Average speed} = \frac{140}{2\tfrac{2}{3}}$$
$$= \frac{420}{8}$$
$$= 52.5 \text{ km h}^{-1}.$$

Example 2. *The distance between Birmingham and Rugby is 48 km.
A bus leaves Birmingham for Rugby at 9.00 a.m. travelling at
60 km h^{-1} and another bus leaves Rugby for Birmingham at 9.15 a.m.
travelling at 50 km h^{-1}. How far are they from Birmingham when
they pass?*

Graphical solution

Take the time axis across the page and the distance axis up the page, starting from Birmingham. The graph representing the journey of the first bus is a straight line through the origin. The graph representing the journey of the second bus is a straight line

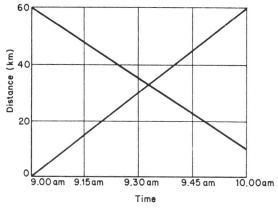

Fig. 1

through the point (9.15 a.m., 48 km). The point of intersection of the two lines in Fig. 1 represents the passing of the buses, 33 km from Birmingham at 9.33 a.m.

Alternative solution

At 9.15 a.m., the bus from Birmingham has travelled 15 km. The distance between the buses when the second bus leaves is therefore 33 km and the buses approach each other at $(60 + 50)$ km h^{-1}. The time taken to close the distance of 33 km is $\frac{33}{110}$ hours, or 18 minutes. The distance travelled in this time by the bus from Birmingham is $\frac{18}{60} \times 60$ km, or 18 km. The distance from Birmingham when they pass is therefore $(15 + 18)$ km, i.e. 33 km.

Example 3. *A particle moves round the circumference of a circle radius 10 cm with constant angular velocity of 8 radians per second. Find the speed of the particle in cm s⁻¹.*

The distance on the circumference subtending an angle of 8 radians at the centre of the circle is 8×10 cm. So the particle travels 80 cm in one second and its speed is 80 cm s⁻¹.

More generally, if a particle is describing a circle radius r cm with an angular velocity of ω radians per second, its speed is $r\omega$ cm s⁻¹. Algebraically, $v = r\omega$.

It will later be necessary to distinguish between angular speed and angular velocity, but the usage here is consistent with common practice at this level.

EXERCISE 1

1. A man runs 1 km in 5 minutes; express his average speed in km h⁻¹.
2. Express a speed of 10^4 cm s⁻¹ in km h⁻¹.
3. A man runs at 12 km h⁻¹. What is his speed in metres per minute?
4. A particle describes the circumference of a circle in 4 seconds. What is its angular velocity in radians per second (rad s⁻¹)?
5. A particle travels round the circumference of a circle radius 10 cm six times per second. Find the average speed of the particle.
6. A car travels 100 km at an average speed of 60 km h⁻¹ and a further 155 km at an average speed of 38.75 km h⁻¹. Find its average speed over the whole journey.
7. If a disc of radius 5 cm is turning about its centre so that any point of the rim is moving at 25 cm s⁻¹, find the angular velocity of the disc in rad s⁻¹.
8. Taking the radius of the Earth to be 6370 km, find the velocity of a point on the equator in km h⁻¹.
9. A man walks 25 km in 4 hours. If he walks the first 10 km at an average speed of 5 km h⁻¹, find his average speed for the remainder of the journey.
10. Taking the distance from the Earth to the Moon to be 380 000 km and the speed of light to be 3×10^8 m s⁻¹, find the time taken for light to travel from the Moon to the Earth.
11. Two towns X and Y are 50 km apart. A bus leaves X at 12 noon to travel to Y at 40 km h⁻¹ and at 12.30 p.m. another bus leaves Y to travel to X at 50 km h⁻¹. How far from X do they meet?
12. The speed of sound is 3.3×10^2 m s⁻¹. Find the time taken to travel 1000 km at a speed of Mach 0.6, which is $0.6 \times$ the speed of sound.

13. Calculate the angular velocity in rad s^{-1} due to rotation about the axis of a point on the Earth's surface in latitude 60°N.

14. A train goes from A to B, a distance of x km, in t hours. It then goes from B to C, a distance of y km in T hours. Find the average speed from A to C.

15. A train travels at u km h^{-1} for t hours and then at v km h^{-1} for T hours. Find its average speed over the whole journey.

16. A train travels x km at u km h^{-1} and a further y km at v km h^{-1}. Find its average speed for the whole journey.

17. A particle travels along a straight track OBA. It goes from O to A, a distance of 40 cm, in 4 seconds and back to B from A, a distance of 10 cm, in 2 seconds. Find
    (a) the average speed,
    (b) the average velocity of the particle.

18. A man walks 2 km north-east and then 2 km north-west in one hour. What is
    (a) his average speed,
    (b) his average velocity?

19. A plane is at a height of 11 000 metres. How long will sound from the plane take to reach an observer immediately under the plane if sound travels at 3.3×10^2 m s^{-1}?

20. Taking the distance between the Earth and the Moon to be 384 000 km, find the time taken for a rocket to reach the Moon if it travels at an average velocity of 12 000 km h^{-1}.

Distance-time graph

One useful way of dealing with problems on velocity is to draw a distance-time graph. The time is plotted along the horizontal axis and the distance along the vertical axis. Suppose that the distances of a particle moving in a straight line from a fixed point of that line are given by the following table:

Distance (metres)	0	2	4	6	8	10
Time (seconds)	0	1	2	3	4	5

When these points are plotted on a distance-time graph as in Fig. 2, we see that all the points lie on a straight line through the origin.

This shows that the motion is consistent with constant velocity and the gradient of the line gives that constant velocity which is 2 m s^{-1}.

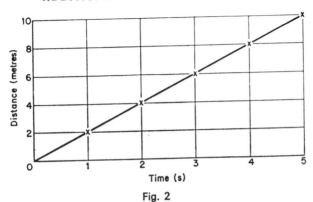

Fig. 2

Variable velocity

In practice, constant velocity is uncommon and velocity generally varies haphazardly with time. If the distance can be expressed algebraically in terms of the time, the problem may be solved by using the methods of the calculus as will be seen later; otherwise the only solution is a graphical one. Suppose corresponding values of distance (s metres) and time (t seconds) are given by the following table:

s	1	4	9	16	25	36
t	1	2	3	4	5	6

The distance-time graph is shown in Fig. 3.

The gradient of the line OP gives the average velocity over the 6 s and is 6 m s^{-1}. The gradient of the line OQ gives the average velocity over the first two seconds and is 2 m s^{-1}. But suppose we wish to find the actual velocity at Q itself? This is certainly not 2 m s^{-1}. It must be greater because the velocity of the particle is increasing all the time. To find the velocity at Q, we consider the average velocity over an interval containing Q and let that interval decrease indefinitely. Suppose that Q_1 and Q_2 are two points on either side of Q and close to Q. The gradient of the chord Q_1Q_2 gives the average velocity over that interval and by choosing Q_1 and Q_2 ever closer to Q, we can get as good an approximation as we wish for the velocity at Q itself. This is, of course, appealing to the methods used in the calculus and we see that the gradient of the tangent at Q gives the actual velocity at Q. The drawing of the tangent by eye is not easy and great accuracy cannot be expected from this method.

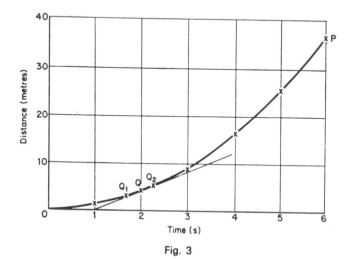

Fig. 3

The tangent at Q passes through the point (1,0) and its gradient is 4, so that the velocity at Q is 4 m s^{-1}.

Velocity at a particular instant therefore is equal to the limiting value of the average velocity over a small interval of time which includes the instant, when that small interval tends to zero. Using the notation of the calculus,

$$v = \frac{\mathrm{d}s}{\mathrm{d}t}.$$

In the example given, the distance and time are connected by the algebraic equation $s = t^2$. From this $\mathrm{d}s/\mathrm{d}t = 2t$ and so the velocity is always given by the expression $2t$, which is the differential coefficient of t^2 with respect to t.

When $t = 2$, $\mathrm{d}s/\mathrm{d}t = 4$ and so the velocity after 2 s is 4 m s^{-1}.

Now let us consider a motion which is not easily expressed

algebraically. Suppose the distances of a train from a station are given in the following table:

Distance (s km)	0	6	17	30	43	53	60
Time	12.00	12.10	12.20	12.30	12.40	12.50	13.00

Distance	55	48	33
Time	13.10	13.20	13.30

The distance-time curve is shown in Fig. 4. Obviously the train soon after 13.00 has reversed and returned towards the station.

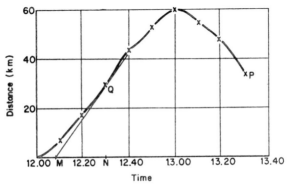

Fig. 4

The average velocity from O to P is the gradient of the line OP which is 33 km/1½ h, i.e. 22 km h^{-1}. The average speed is greater than this because the total distance travelled is certainly greater than 60 km. So it is essential to distinguish between average velocity and average speed.

$$\text{Average velocity} = \frac{\text{total displacement}}{\text{total time}}.$$

$$\text{Average speed} = \frac{\text{total distance travelled}}{\text{total time}}.$$

The velocity at 12.30, say, may be estimated by drawing by eye the tangent at that instant. The velocity is given by the gradient QN/MN and is equal to 30 km/22 min, approximately 82 km h^{-1}.

Notice that the information was insufficient to draw the graph accurately. We deduced that the train turned round but we are not told how long this operation took although from the figures it is obvious that not much time was wasted.

Example 1. *The height (s metres) of a stone thrown vertically upwards above the ground is given in terms of the time (t seconds) by* $s = 70t - 4.9t^2$. Calculate
 (a) the initial velocity of the stone,
 (b) the greatest height reached by the stone,
 (c) the time taken before the stone again reaches the ground.

(a) $s = 70t - 4.9t^2$

$\Rightarrow v = \mathrm{d}s/\mathrm{d}t = 70 - 9.8t.$

When $t = 0$, $v = 70$, i.e. the initial velocity is 70 m s^{-1}.

(b) At the greatest height, the velocity is momentarily zero.

$$\mathrm{d}s/\mathrm{d}t = 0$$
$$\Rightarrow 70 = 9.8t$$
$$t = \tfrac{50}{7}$$
$$= 7\tfrac{1}{7}.$$

When $t = 7\tfrac{1}{7}$,

$$s = 70t - 4.9t^2$$
$$= 70(\tfrac{50}{7}) - 4.9(\tfrac{2500}{49})$$
$$= 500 - 250$$
$$= 250.$$

The greatest height is 250 m.

(c) When the stone reaches the ground again, the height of the stone is zero.

$$70t - 4.9t^2 = 0$$
$$\Rightarrow t = 0 \quad \text{or} \quad t = \tfrac{100}{7}$$

Thus
$$t = 14\tfrac{2}{7}, \text{ to reach the ground again.}$$

This as we should expect is double the time to the highest point.

Example 2. *The velocity (v metres per second) of a particle moving along a straight line is given in terms of the time (t seconds) by the equation v = 20 + 4t. Calculate*

    (*a*) *the initial velocity of the particle,*

    (*b*) *the distance gone in the first 5 seconds,*

    (*c*) *the average velocity during the first 5 seconds,*

    (*d*) *the velocity after 5 seconds.*

(a) The initial velocity is found by putting $t = 0$ in $(20 + 4t)$ and is 20 m s^{-1}.

(b) Since $v = \mathrm{d}s/\mathrm{d}t$, it follows that $s = \int v \, \mathrm{d}t$.

$$\therefore \quad s = \int (20 + 4t) \, \mathrm{d}t$$

$$= 20t + 2t^2 + C.$$

When $t = 0$, $s = 0$ and so $C = 0$.

$$\therefore \quad s = 20t + 2t^2.$$

When $t = 5$, $s = 100 + 50 = 150$, i.e. the distance gone in the first 5 s is 150 m.

(c) Average velocity $= \dfrac{\text{total displacement}}{\text{total time}}$

$$= \frac{150 \text{ m}}{5 \text{ s}} = 30 \text{ m s}^{-1}.$$

(d) Since $v = 20 + 4t$, therefore $v = 40$ when t = 5. The velocity after 5 s is 40 m s^{-1}.

EXERCISE 2

1. Taking the distance of the Earth from the Sun to be 1.5×10^8 km and the length of the year to be 365 days, find the speed of the Earth in its orbit, in km h^{-1}.

2. The times taken by a runner to pass certain distance posts were:

s (metres)	25	60	80	110	135	160	185	200
t (seconds)	3.5	6	8	11.5	14.5	18	21.5	25

Draw a distance-time graph. From your graph, estimate his speed when he has been running 7 seconds.

3. A car during a rally test has to travel 100 m forward along a straight road, reverse for 40 m and then go forward again another 100 m. The times for the three phases are 4 s, 3 s and 5 s. Draw a distance-time curve and calculate
   (a) the average speed,
   (b) the average velocity.

4. Express a velocity of 180 cm s^{-1} in km h^{-1}.

5. Express a velocity of 30 km h^{-1} in cm s^{-1}.

6. If the relation between the distance (s metres) and the time (t seconds) is given by $s = 4t^2$, plot the distance-time curve and estimate the velocity when $t = 2$.

7. Plot the following values of s and t:

s (metres)	3	4	7	12	19	28	39
t (seconds)	0	1	2	3	4	5	6

Find the velocity when $t = 1, 2, 3, 4, 5, 6$. What do you notice?

8. Plot the following values of s and t:

s (metres)	0	3	8	15	24	35
t (seconds)	0	1	2	3	4	5

Find the velocity when $t = 1, 2, 3, 4, 5$.

9. A body falls a distance h metres in t seconds from rest according to the law $h = 4.9t^2$.
   (a) How far does it fall in 3 seconds?
   (b) How far does it fall in the third second?
   (c) What is its velocity after 3 seconds?

10. A ball thrown vertically upwards from a window falls to the ground below. The height (h metres) above the ground is given in terms of the time (t seconds) after it is thrown by the equation

$$h = 20 + 47t - 4.9t^2.$$

   (a) How high is the window above the ground?
   (b) With what velocity is the ball thrown?
   (c) How long does it take to reach the ground?

11. The velocity, v m s^{-1} of a body travelling in a straight line t seconds after passing a fixed point O of the line is given by

$$v = 3 + 4t.$$

Find the distance from O when $t = 2$.

12. The velocity of a body (v m s^{-1}) travelling in a straight line t seconds after passing a point O of the line is given by the table:

v	4	7	10	13	16
t	0	1	2	3	4

What is the increase of velocity in each second?
What would you expect the velocity to be after 5 seconds?

13. If a body starting from rest increases its velocity by 4 cm s^{-1} in every second, what is its velocity after 5 s?
What is its average velocity during this time?

14. The angular velocity of a disc is 100 rad s^{-1}. Express this in revolutions per minute.

15. The angular velocity of a wheel is 1000 revolutions per minute. If the radius of the wheel is 10 cm, find the velocity of a point on its circumference.

16. If the distance (s metres) travelled in t seconds by a particle moving in a straight line is given by $s = 20t - 10t^2$, find
    (a) the velocity when $t = 0$,
    (b) the time when the particle returns to its starting point.

17. If the distance (s metres) travelled in t seconds by a particle moving in a straight line is given by $s = ut + 4t^2$, where u is a constant, find the initial velocity of the particle.

18. If the velocity, v m s^{-1}, of a particle travelling in a straight line is given by $v = 2t^2$, where t is the time in seconds, find the distance gone in the first 3 seconds.

19. A train goes from station A to station B in one minute. Its velocity v metres per second after t seconds is given by:

$$v = 0.8t \qquad (t < 10)$$
$$v = 8 \qquad (10 \leqslant t \leqslant 50)$$
$$v = 48 - 0.8t \qquad (50 < t \leqslant 60).$$

Find the distance between the stations.

20. If a body increases its velocity at a constant rate of 4 m s^{-1} in every second, its acceleration is said to be 4 m s^{-2}. Given that the velocity, v m s^{-1} is given in terms of the time t seconds by the equation
$$v = 4 + 3t$$
find the acceleration.

CHAPTER 2

Acceleration

Uniform acceleration

Suppose that the velocity of a body increases with the time as in the following table:

Velocity ($m\ s^{-1}$)	0	4	8	12	16	20
Time (s)	0	1	2	3	4	5

The increase in velocity in each second is constant and is $4\ m\ s^{-1}$. So the velocity is increasing at the constant rate of $4\ m\ s^{-1}$ per second or $4\ m\ s^{-2}$. The *acceleration* of the body is said to be $4\ m\ s^{-2}$. If we assume that the rate of increase of the velocity is uniform between the times given (which it need not be) then the body has a constant acceleration of $4\ m\ s^{-2}$. If the acceleration is negative, it is called a *retardation*.

Velocity-time graph

If the velocity in the example given is plotted against the time, the graph is a straight line as shown in Fig. 5.

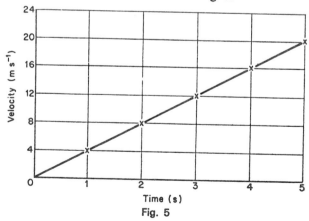

Fig. 5

N.B. The time axis should always be horizontal.

The gradient of the line which is 20 m s^{-1}/5 s or 4 m s^{-2} gives the acceleration. Whenever the velocity of a body is changing, that body must have an acceleration and generally that acceleration will not be constant. The acceleration is always given by the gradient of the tangent to the velocity-time graph at the appropriate instant. The tangent may be drawn by eye but this method is not very accurate.

Using the notation of the calculus and denoting the acceleration by a,

$$a = \frac{\mathrm{d}v}{\mathrm{d}t}$$

and since $v = \mathrm{d}s/\mathrm{d}t$, it follows that

$$a = \frac{\mathrm{d}^2 s}{\mathrm{d}t^2}.$$

Just as the velocity is the rate at which distance is increasing with time, the acceleration is the rate at which velocity is increasing with time.

The *average acceleration* over an interval of time is equal to

$$\frac{\text{increase of velocity during that interval}}{\text{time taken}}.$$

To find the acceleration at a particular instant, we consider the average acceleration over an interval containing that instant and let the time interval diminish indefinitely.

Example 1. *A body is moving with uniform acceleration. If its velocity after 2 s is 10 m s^{-1} and after 5 s is 19 m s^{-1}, find its acceleration.*

Increase of velocity in 3 s is 9 m s^{-1}. Acceleration is

$$\frac{9 \text{ m s}^{-1}}{3 \text{ s}} = 3 \text{ m s}^{-2}.$$

Example 2. *A body has a uniform acceleration of 8 cm s^{-2}. If its initial velocity is 20 cm s^{-1}, find its velocity after 4 s.*

The increase of velocity is 8 cm s^{-1} per second.

The increase of velocity in 4 seconds is 32 cm s^{-1}.

Its velocity is 52 cm s^{-1}.

Area under the velocity-time curve

Suppose that Fig. 6 shows a velocity-time curve.

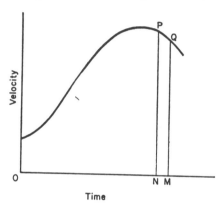

Fig. 6

Let PN and QM be ordinates drawn from two 'adjacent' points P and Q on the curve. The area of the figure PQMN may be taken as the area of a rectangle if P is close to Q and the area of this figure is approximately PN.NM or the product of the velocity at P and the time from P to Q. Since this time is very small, we may take the velocity as constant and this product therefore equals the distance gone. By dividing the area under the curve up into countless such rectangles, we see that the area under the curve and between any two ordinates is equal to the distance travelled in the time between the ordinates.

For a more formal proof using calculus, the area under a curve is equal to $\int y \, dx$. In this case, the x-axis is the time-axis and the

y-axis is the velocity-axis, so the area under the curve equals $\int v\,dt$. But $v = ds/dt$,

$$\therefore \quad A = \int \frac{ds}{dt}\,dt$$

$$= \int ds$$

$$= \text{distance gone between the appropriate ordinates.}$$

This area may be found by counting squares, by the trapezoidal rule or by Simpson's rule (see *Additional Pure Mathematics*, p. 148). When finding the distance by counting squares, difficulty is often experienced with the units. If 1 cm on the time axis represents 10 s and 1 cm on the velocity axis represents 5 m s^{-1}, then 1 cm^2 represents the distance travelled in 10 s with a velocity of 5 m s^{-1}, i.e. 50 m.

If the graph goes below the time axis, the velocity during that part of the motion will be negative. The displacement during that time is also negative so that to find the total displacement (as distinct from the total distance travelled), the area above the time axis must be taken as positive, the area below as negative.

Example (*Exercise 2, question 19*). *A train goes from station A to station B in one minute. Its velocity v metres per second after t seconds is given by*

$$v = 0.8t \qquad (t < 10)$$
$$v = 8 \qquad\quad (10 \leqslant t \leqslant 50)$$
$$v = 48 - 0.8t \quad (50 < t \leqslant 60).$$

Find the distance between the stations.

The velocity-time graph consists of three straight lines as shown in Fig. 7.

The distance between the stations is given by the area of the trapezium. The area of each triangle is

$$\tfrac{1}{2}(8 \text{ m s}^{-1})(10 \text{ s}) = 40 \text{ m}.$$

The area of the rectangle is

$$(8 \text{ m s}^{-1})(40 \text{ s}) = 320 \text{ m}.$$

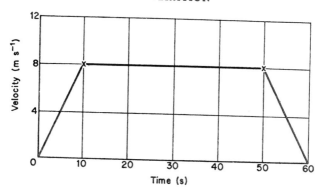

Fig. 7

The total distance gone is

$$320 \text{ m} + 40 \text{ m} + 40 \text{ m} = 400 \text{ m}.$$

EXERCISE 3

1. A train increases its speed from 40 km h^{-1} to 50 km h^{-1}. If this takes 30 s, find its average acceleration in km h^{-2}.
2. A car brakes from 45 km h^{-1} to 30 km h^{-1} in 4 s. What is its average retardation in m s^{-2}?
3. Express an acceleration of 1000 km h^{-2} in m s^{-2}.
4. Express a retardation of 5 m s^{-2} in km h^{-2}.
5. The velocity ($v \text{ m s}^{-1}$) of a car in terms of the time (t seconds) is shown in the following table:

v	4.0	4.1	4.4	4.9	5.6	6.5	8.9
t	0	1	2	3	4	5	6

Find the acceleration when $t = 3$.

6. The speed of a train ($v \text{ m s}^{-1}$) and the time (t seconds) are given in the table:

v	0	1.5	4.0	7.5	12.0	17.5	23.0
t	0	1	2	3	4	5	6

Find the distance travelled in the first 6 seconds.

7. Between two stations a train accelerates uniformly at first, then moves with constant speed and finally retards uniformly. If the ratios of the times taken are 1 : 6 : 1, and the greatest speed is 40 km h^{-1}, find the average speed over the whole journey.

8. The acceleration of a body is 9.8 m s^{-2}. If the velocity of the body after 4 s is 58 m s^{-1}, calculate its initial velocity.

9. A body moves with uniform acceleration. If its velocity after 4 s is 8 m s^{-1} and after 10 s is 56 m s^{-1}, find its acceleration.

10. If the velocity of a train is reduced from 15 m s^{-1} to 10 m s^{-1} in 1 minute, what is its average retardation?

11. A train takes 5 minutes to go from one station to another, 2 km away. It first accelerates uniformly, then runs at maximum speed for 3 minutes and finally retards uniformly with a retardation numerically equal to the acceleration. Find its maximum speed.

12. A train takes 6 minutes to go from one station to another. It accelerates uniformly for 1 minute, runs at a constant speed of 40 km h^{-1} for 4 minutes and finally retards uniformly for 1 minute. Find the distance between the stations.

13. The number of revolutions per second (r.p.s.) of a flywheel slowing down is given in terms of the time in the following table:

r.p.s.	360	351	324	279	216	135
Time (seconds)	0	30	60	90	120	150

Find the retardation in revolutions per second per second after 100 s.

14. The speed of a train between two stations is given in the following table:

Time from start (minutes)	1	2	3	4	5	6
Speed (km h^{-1})	20	40	60	60	30	0

Assuming that acceleration and retardation are uniform, plot a velocity-time curve.
Find the distance between the stations.

15. A flywheel rolls down a pair of rails with constant acceleration. The distance moved by the centre of the flywheel in terms of the time is given in the table:

Distance (cm)	0	10	40	90
Time (seconds)	0	1	2	3

(a) When is the velocity of the centre of the flywheel 20 cm s^{-1}?
(b) What is the acceleration of the centre of the flywheel?

16. The distances moved by a body which has constant acceleration are given in the table:

s (cm)	0	25	60	105	160	225
t (seconds)	0	1	2	3	4	5

Find its acceleration and initial velocity.

17. A train starts from rest with an acceleration of a m s^{-2}. Its acceleration during the first 60 s of its motion is $(a - \frac{1}{2}t)$ m s^{-2} where t seconds is the time after the start. If the train has a speed of 36 km h^{-1} at the end of the first 60 s, find the value of a.

18. The front of a train which is being uniformly retarded passes a certain point at 60 km h^{-1} and 3 s later the tail of the train passes the same point at 54 km h^{-1}. Find the length of the train.

19. Between two stations 300 m apart a train is first uniformly accelerated at 1 m s^{-2} until it attains its maximum speed and then immediately retarded at 2 m s^{-2} until it comes to rest at the second station. Find the maximum speed of the train in km h^{-1}.

20. A body moving with uniform acceleration in a straight line travels a distance of 6 m in one second and a distance of 8 m in the following second. Find the velocity of the body after 0.5 s and after 1.5 s and hence its acceleration.

Motion under uniform acceleration

Suppose that a body has initial velocity u and that its velocity after time t is v; also that the uniform acceleration is a and the distance gone in time t is s. The velocity-time graph is shown in Fig. 8. As the acceleration is uniform, the graph is a straight line.

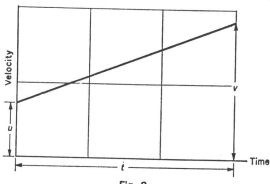

Fig. 8

The gradient of the line gives the acceleration.

$$\text{The gradient is } \frac{v-u}{t}, \text{ i.e. } a = \frac{v-u}{t} \qquad \text{(i)}$$

$$\Rightarrow v = u + at.$$

The area of the trapezium is the distance travelled in time t.
The area of a trapezium = $\frac{1}{2}$ (sum of the parallel sides) × distance between them.

$$\therefore \quad s = \frac{1}{2}(u+v)t. \qquad \text{(ii)}$$

From (i), $v - u = at.$

From (ii), $v + u = \dfrac{2s}{t}$

$$\therefore \quad (v-u)(v+u) = at\left[\frac{2s}{t}\right]$$

$$\Rightarrow v^2 = u^2 + 2as. \qquad \text{(iii)}$$

Substituting $v = u + at$ in (ii), we have

$$s = \frac{u + u + at}{2} t$$

$$= ut + \tfrac{1}{2}at^2. \qquad \text{(iv)}$$

Substituting $u = v - at$ in (ii), we have

$$s = \frac{v - at + v}{2} t$$

$$= vt - \tfrac{1}{2}at^2. \qquad \text{(v)}$$

N.B. These five equations may only be used when the acceleration is constant and motion is in the same straight line.

We have five variables s, a, t, v, u and the equations (i) to (v), omit each of these variables in turn. For example, equation (i) contains all the variables except s, equation (v) contains all the variables except u.

Use of calculus

Let V be the velocity after any time T. Then

$$\frac{dV}{dT} = a \qquad \text{(a constant)}$$

$$\therefore \quad V = \int a\, dT$$

$$= aT + C.$$

When $T = 0$, $V = u$ and so $C = u$. Hence

$$V = aT + u.$$

When $T = t$, $V = v$. Hence

$$v = u + at.$$

Since $V = u + aT$ and $V = dS/dT$ where S is the distance gone in time T,

$$\frac{dS}{dT} = u + aT$$

$$\Rightarrow S = uT + \tfrac{1}{2}aT^2 + K.$$

When $T = 0$, $S = 0$. Therefore $K = 0$, and

$$S = uT + \tfrac{1}{2}aT^2.$$

When $T = t$, $S = s$ and so

$$s = ut + \tfrac{1}{2}at^2.$$

From $s = ut + \tfrac{1}{2}at^2$, $s = t(u + at) - \tfrac{1}{2}at^2$ we get

$$s = vt - \tfrac{1}{2}at^2.$$

Squaring each side of $v = u + at$,

$$v^2 = u^2 + 2uat + a^2t^2$$

$$\Rightarrow v^2 = u^2 + 2a(ut + \tfrac{1}{2}at^2)$$

$$\Rightarrow v^2 = u^2 + 2as.$$

Since $v^2 - u^2 = 2as$ and $v - u = at$,

$$\frac{v^2 - u^2}{v - u} = \frac{2as}{at}$$

$$\Rightarrow v + u = \frac{2s}{t}$$

$$\Rightarrow s = \frac{u + v}{2}\, t.$$

It is worth noting that from $v^2 = u^2 + 2as$, we get by differentiating with respect to s,

$$2v\,\frac{\mathrm{d}v}{\mathrm{d}s} = 2a,$$

i.e., $$a = v\,\frac{\mathrm{d}v}{\mathrm{d}s}.$$

This equation may also be found independently from $a = \mathrm{d}v/\mathrm{d}t$.

$$a = \frac{\mathrm{d}v}{\mathrm{d}t} = \frac{\mathrm{d}v}{\mathrm{d}s}\frac{\mathrm{d}s}{\mathrm{d}t} = v\,\frac{\mathrm{d}v}{\mathrm{d}s}$$

since $v = \mathrm{d}s/\mathrm{d}t$. This equation is useful when the acceleration or velocity is given in terms of the distance and not the time.

Notice also that by integrating $a = v\,\mathrm{d}v/\mathrm{d}s$ with respect to s, the equation $v^2 = u^2 + 2as$ may be deduced.

The equation $$s = \frac{u + v}{2}\, t$$

$$\Rightarrow \frac{s}{t} = \frac{u + v}{2}.$$

So for uniformly accelerated motion in a straight line, the average velocity which is given by s/t is equal to the mean of the initial and final velocities.

When the acceleration is not constant, calculus methods must be used and an example is given below.

Example 1. *If a train running at 40 km h^{-1} is retarded uniformly to stop in 4 minutes, find the distance it moves before it comes to rest.*

We must be careful to work in the same units throughout. Let us choose km and hours.

Then $u = 40$, $v = 0$, $t = \frac{1}{15}$. We wish to find s. The equation connecting these four quantities is

$$s = \frac{u + v}{2} t = \frac{40}{2} \left[\frac{1}{15} \right] = 1\tfrac{1}{3}.$$

The distance is $1\tfrac{1}{3}$ km.

Example 2. *A particle whose initial velocity is 10 cm s^{-1} moves for 6 seconds in a straight line with an acceleration of 4 cm s^{-2} and then with retardation of 8 cm s^{-2}. How far does it go before it comes to rest?*

The two parts of the motion must be treated separately. Using centimetres and seconds, for the accelerated motion we are given that $u = 10$, $t = 6$ and $a = 4$.

$$v = u + at$$
$$\Rightarrow v = 10 + 24$$
$$= 34.$$

Also
$$s = ut + \tfrac{1}{2}at^2$$
$$\Rightarrow s = 60 + 2(36)$$
$$= 132.$$

The second part of the motion therefore starts with a velocity of 34 cm s^{-1}. So $u = 34$, $v = 0$, $a = -8$.

$$v^2 = u^2 + 2as$$
$$\Rightarrow 0 = 34^2 - 16s$$
$$\Rightarrow s = \frac{34 \times 34}{16}$$
$$= \frac{289}{4} = 72\tfrac{1}{4}.$$

The total distance gone is $(132 + 72\tfrac{1}{4})$ cm or $204\tfrac{1}{4}$ cm.

Example 3. *The velocity of a body moving in a straight line is given in terms of the distance by the equation* $v^2 = 64 - 4s^2$. *Show that the retardation is proportional to the distance.*

Differentiating $v^2 = 64 - 4s^2$ with respect to s, we get

$$2v \frac{dv}{ds} = -8s$$

$$\Rightarrow a = v \frac{dv}{ds} = -4s.$$

Since a is negative when s is positive, this equation shows that the retardation is proportional to the distance travelled.

Example 4. *If the distance (s cm) travelled by a body moving in a straight line is connected with the time (t seconds) by the equation $s = 4t - t^3$, find (i) its initial velocity, (ii) its acceleration after 2 s.*

The acceleration is not constant so none of the constant acceleration formulae may be used.

$$s = 4t - t^3$$

$$\Rightarrow v = \frac{ds}{dt} = 4 - 3t^2$$

$$\Rightarrow a = \frac{dv}{dt} = -6t.$$

(i) From $v = 4 - 3t^2$, when $t = 0$, $v = 4$. The initial velocity is 4 cm s^{-1}.

(ii) From $a = -6t$, when $t = 2$, $a = -12$. The retardation is 12 cm s^{-2}.

EXERCISE 4

1. If a car is running at 30 km h^{-1} and is stopped in 60 m, find the time taken to stop it.
2. If a train running at 48 km h^{-1} is stopped in 4 minutes, find the distance travelled before it stops.
3. Assuming that acceleration is constant, if a shell acquires a speed of 300 m s^{-1} after travelling 6 m along the barrel of a gun, find the time taken.

4. A stone falling with an acceleration of 9.8 m s^{-2} starts from rest. Find its speed when it has travelled 40 m.

5. A car travelling with uniform acceleration clocks speeds of 7 m s^{-1} and 13 m s^{-1} at an interval of 6 s. Find its acceleration.

6. A car accelerates at 2 m s^{-2}. Its speed initially is 10 m s^{-1}. Find the distance travelled in the next 5 s.

7. What retardation in km h^{-2} is needed to decrease the speed of a car from 30 km h^{-1} to 10 km h^{-1} in (i) 10 seconds, (ii) 100 metres?

8. A train accelerating at 0.1 m s^{-2} passes a signal box at 60 km h^{-1}. How far did it travel in the previous minute?

9. A car increases its speed from 5 km h^{-1} to 35 km h^{-1} in a distance of 200 m. Find its acceleration in km h^{-2}.

10. The front of a train which is being uniformly retarded passes a signal box at 60 km h^{-1} and 4 s later the tail of the train passes the signal box at 36 km h^{-1}. Find the length of the train in metres.

11. A stone thrown across ice has a retardation of 0.5 m s^{-2}. How far will it travel if its initial speed is 10 m s^{-1}?

12. The distance, s metres, travelled by a car in the time, t seconds, from the moment the brakes are applied is given by the formula

$$s = 10 \left(t - \tfrac{1}{48}t^3\right).$$

Find the time taken for the car to stop.

13. In question 12, find the distance travelled before the car comes to rest.

14. A trolley rolls down an inclined plane with a constant acceleration of 2 m s^{-2}. Find the distance travelled between the instants when its speeds are 10 m s^{-1} and 16 m s^{-1}.

15. A coach travelling at 45 km h^{-1} pulls up in 100 m. What is the retardation in m s^{-2}?

16. A body moving with uniform acceleration in a straight line travels a distance of 6 m in one second and a distance of 10 m in the following second. Find the acceleration of the body.

17. A cyclist reaches the top of a straight hill at 5 m s^{-1} and accelerates uniformly down the hill to reach the bottom at 15 m s^{-1}. If the hill is 50 m long, find his acceleration.

18. The distance s metres travelled by a body moving in a straight line in t seconds is given by $s = 3t^3 - 4t^2$. Find the velocity after 2 s and the initial acceleration.

19. The distance, s metres, moved by a particle travelling in a straight line in t seconds is given by the equation $s = 2t + t^3$. Calculate
   (i) the average velocity over the first 3 s;
   (ii) the velocity after 3 s.

20. The distance, s metres, travelled by a body in time, t seconds, satisfies the equation $s = t^3 - 2t$. Find the acceleration after 2 seconds.

21. The velocity of a body is given in terms of the distance s by the equation $v^2 = 4s$. Show that the acceleration is constant.

22. The acceleration of a small rocket in m s^{-2} is numerically equal to the distance travelled in metres. If the initial velocity of the rocket is 3 m s^{-1}, find its velocity when it has travelled 4 m.

23. The acceleration a m s^{-2}, of a body moving in a straight line is given in terms of the time, t seconds, by the equation $a = 3 + 4t$. If its initial velocity is 10 m s^{-1}, find its velocity after 2 s.

24. The velocity, v m s^{-1}, of a train is given in terms of the time, t seconds, for the first minute of its motion by the equation $1000v = 40t + t^2$. Find the distance travelled in the first minute.

25. A motorist travels x km at an average speed of u km h^{-1}, rests for t minutes and then travels a further y km at an average speed of v km h^{-1}. Find his average speed for the whole journey in km h^{-1}.

CHAPTER 3

Mass, Weight

Mass as a fundamental quantity

What makes a stone fall to the ground if it is thrown up? We all know the answer to that one. It is the pull of the earth, or the gravitational pull, on the stone. The size of this pull depends upon the stone; upon some indefinable property of the stone which is not its size but which depends also on the substance of which the stone is made. This indefinable property is called the *mass* of the stone and cannot be defined because it is one of the fundamentals like time and length, in terms of which every other quantity is defined. Try to define time and you will find out the difficulty for yourself. You can do it only by using words such as interval which basically comes down to saying that 'time is time'. This third basic quantity is called the mass of a body; the unit of mass is the kilogramme (kg) or the gramme (g), where 1 kg = 1000 g. The force with which the earth attracts a body is proportional to the mass of the body, e.g. the attraction on a mass of 2 kg is double that on a mass of 1 kg. Since it is a force, it is measured in newtons (N).

In these space days, we know from television programmes and television broadcasts that this is true within certain bounds only; that it is indeed possible to travel outside the earth's gravitational field; but we are going now to consider motion for relatively small heights above the earth's surface and then, for any given place, the gravitational pull is proportional to the mass.

Acceleration due to gravity

We have said that for a larger mass the gravitational pull is larger but, as Galileo first discovered by conducting experiments from the leaning tower of Pisa, this is not true of the acceleration produced. In results published in 1638, Galileo stated that a body falling under gravity falls with a constant acceleration and does not (provided that we neglect the effect of air resistance) depend on the size, shape or mass of the body. This acceleration is called the acceleration due to gravity and is denoted by g. The value of g throughout this book is taken as 9.8 m s^{-2}. It should be remembered that this value is true to 2 significant figures only and that results derived from the figure 9.8 can only be true to that number of significant figures. The value of g does vary slightly over the surface of the earth and is largest at the poles 9.832 m s^{-2}, least at the equator 9.780 m s^{-2}. At Greenwich, its value is 9.812 m s^{-2} to four significant figures, but for convenience we shall always use the value 9.80 in our calculations.

The five equations for uniform acceleration proved in the last chapter hold for motion under gravity with a taking the value 9.8 m s^{-2} downwards. Remember when using these equations that velocity may be upwards or downwards and that distance may be upwards or downwards. So in all problems connected with gravity, take the upward (or downward if more convenient) direction to be positive; a downwards velocity must then be given a negative value.

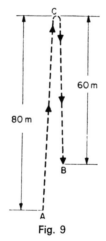

Fig. 9

If we are considering the motion of a stone from A to B as shown in Fig. 9, the distance from A to C is $+80$ m and the distance from C to B is -60 m.

The distance moved from A to B is therefore +20 m or, in other words, the value of s inserted in the formulae is the change of height with the appropriate sign.

Suppose that a rocket is projected vertically upwards with a velocity of 70 m s^{-1}. It will lose 9.8 m s^{-1} in velocity every second, and its velocity at the end of each successive second will be 60.2 m s^{-1}, 50.4 m s^{-1}, 40.6 m s^{-1} and so on. These numbers form an arithmetic progression with common difference -9.8. What will be the velocity after 10 s? It will be $(70 - 98)$m s^{-1} or -28 m s^{-1}. This means that it will be moving downwards with a velocity of 28 m s^{-1}. So we see that there is no need to consider the upward and downward motions separately.

What is the distance gone in the first second? The initial and final velocities are 70 m s^{-1} and 60.2 m s^{-1}, so that the distance gone is $\frac{1}{2}(70 + 60.2)$ m or 65.1 m. In the next second, the distance gone is $\frac{1}{2}(60.2 + 50.4)$ m or 55.3 m. In the third second, the distance gone is $\frac{1}{2}(50.4 + 40.6)$ m or 45.5 m. The distances travelled in consecutive seconds are therefore 65.1 m, 55.3 m, 45.5 m . . . . These numbers also form an arithmetic progression with common difference -9.8.

What is the distance gone in 14 s? Using $s = ut + \frac{1}{2}at^2$ and putting $u = 70$, $t = 14$ and $a = -9.8$, we have

$$s = 70(14) - \frac{1}{2}(9.8)14^2 = 19.6.$$

This means that the rocket is now 19.6 m above the ground and not that the total distance travelled is 19.6 m.

How long does the rocket take to reach its highest point? Since the velocity decreases at the rate of 9.8 m s^{-1} in every sec, the time is

$$\frac{70}{9.8} \text{ s} \quad \text{or} \quad 7\frac{1}{7} \text{ s}.$$

What is the greatest height? Using the equation $s = \frac{1}{2}(u + v)t$ and putting $u = 70$, $v = 0$ (since the velocity at the greatest height is zero), $t = 7\frac{1}{7}$, we have

$$s = \frac{70}{2} \left(\frac{50}{7}\right) \text{ m} = 250 \text{ m}.$$

What is the time taken to reach the ground again? When the rocket reaches the ground again, the change in height from the beginning of the motion is zero.

Use the equation $s = ut + \frac{1}{2}at^2$, putting $s = 0$, $u = 70$ and $a = -9.8$. Then

$$0 = 70t - 4.9t^2$$

$$\Rightarrow t = 0 \quad \text{or} \quad \frac{70}{4.9}.$$

Hence the time taken is $14\frac{2}{7}$s. This, as we would expect, shows that the time taken to reach the ground again is double the time to the highest point.

Weight

The weight of a man is the force with which the earth attracts him. In these days of space travel, we know that as man recedes from the earth his weight becomes less and finally that in space he becomes weightless. So weight is a variable quantity. It varies slightly over the surface of the earth due to the change in g and changes much more radically as he moves into space. Mass, on the other hand, is constant and invariable and the mass of a man in space is exactly the same as his mass on earth. For the purposes of this book, weight is taken to be the force of attraction of the earth when g is 9.8 m s^{-2}.

Example 1. *A stone is thrown vertically upwards with a velocity of 20 m s^{-1}. Find*

   (i)   its greatest height,
   (ii) the time taken to reach the highest point,
   (iii) the velocity when it reaches the ground again.

(i) Use $v^2 = u^2 + 2as$ and put $v = 0$, $u = 20$, $a = -9.8$.

$$0 = 400 - 19.6s$$

$$\Rightarrow s = \frac{400}{19.6} = 20.4, \text{ approximately.}$$

The greatest height is 20.4 m.

(ii) Use $s = ut + \frac{1}{2}at^2$ and put $s = 0$, $u = 20$, $a = -9.8$.

$$0 = 20t - 4.9t^2$$

$$\Rightarrow t = 0 \quad \text{or} \quad \frac{20}{4.9}$$

The time taken to reach the ground again is 20/4.9 s, about 4.08 s. The time taken to reach the highest point is 2.04 s.

(iii) Use $v^2 = u^2 + 2as$ for the whole flight. Then $s = 0$ and so $v^2 = u^2$. At the beginning of the flight $v = u$ and at the end of the flight $v = -u$.

The final speed is equal to the initial speed and is 20 m s^{-1}.

Example 2. *A stone is thrown vertically upwards with a velocity of 4.9 m s^{-1} from a window 29.4 m above the ground. How long does it take to reach the ground?*

If we take the upward direction as positive, $u = 4.9$, $s = -29.4$ and $a = -9.8$.

$$s = ut + \tfrac{1}{2}at^2$$
$$\Rightarrow -29.4 = 4.9t - 4.9t^2$$
$$4.9t^2 - 4.9t - 29.4 = 0$$
$$7t^2 - 7t - 42 = 0$$
$$t^2 - t - 6 = 0$$
$$(t - 3)(t + 2) = 0$$
$$\Rightarrow t = 3 \quad \text{or} \quad t = -2.$$

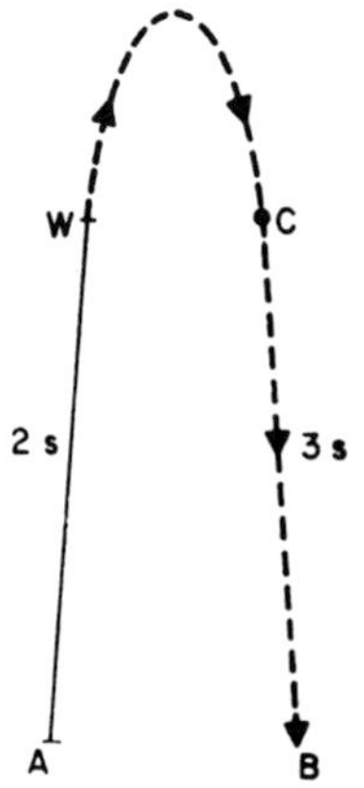

Fig. 10

Obviously the time taken to reach the ground is 3 s but what does the rejected $t = -2$ mean? The equation of motion has not been able to take account of the fact that the motion begins at the window and has given us information about a complete flight from ground to ground. So $t = -2$ means that 2 s before it was at window level, it would have been at ground level (if the motion had begun there).

Fig. 10 shows that it takes 2 s to travel from A to W and 3 s to travel from W to B. It therefore by symmetry takes 2 s to travel from C to B and so 1 s to travel from W to C.
The time taken to the highest point is therefore 0.5 s.

EXERCISE 5

1. A cup falls from a table 1 m above the floor. How long does it take to reach the floor?
2. A stone dropped from a bridge reaches the ground in 2 s. How high is the bridge?
3. A cricket ball thrown vertically upwards takes 4 s to reach the ground again. How high does it rise?
4. A balloon which is stationary starts to rise with an acceleration of 2 m s^{-2}. What is its velocity 10 s later?
   If ballast is dropped at the end of 10 s, what will be the velocity of the ballast after another 10 s?
5. A rocket is projected vertically upwards with a velocity of 24.5 m s^{-1}. For what length of time is it more than 29.4 m above the ground?
6. A parachutist reaches the ground with a velocity of 5 m s^{-1}. From what height would he have to fall freely to gain the same velocity?
7. A stone dropped down a well reaches the bottom in 4 s. How deep is the well?
8. A window is 14.1 m above the ground. A boy throws a stone vertically upwards from the window with a velocity of 10 m s^{-1}. How long does it take to reach the ground?
9. A trolley running down an inclined plane has an acceleration of $\frac{1}{2}g$. How long does it take to travel 39.2 m from rest?
10. An aeroplane diving has an acceleration of $2g$. What additional velocity does it acquire in 10 s?
11. A stone is thrown upwards from the foot of a cliff with a velocity of 20 m s^{-1} at the same instant that a stone is let fall from the top of a cliff immediately above the first stone. They meet after 2 s. What is the height of the cliff?

12. A stone is thrown upwards with a velocity of V m s^{-1} at the same instant that a stone is let fall from the top of a cliff immediately above the first stone. They meet after t s. What is the height of the cliff?

13. A rocket is projected vertically upwards with a velocity of v m s^{-1}. Find:
 (i) the greatest height reached,
 (ii) the time taken to reach the ground again.

14. A stone is dropped from a stationary balloon. Find the distance travelled by the stone in the fifth second of its motion.

15. A sandbag dropped from a stationary balloon travels 34.3 m in the time between t s and $(t + 1)$ s after it is dropped. Find the numerical value of t.

16. A fire hose delivers water vertically upwards with a velocity of 20 m s^{-1}. How high does the jet reach?

17. Two stones are thrown vertically upwards, the first with a velocity of 40 m s^{-1} and the other, 2 s later, with a velocity of 60 m s^{-1}. How long after the second stone is thrown are the stones at the same height?

18. A ball is dropped from a height of 5 m on to a stone floor. If it rebounds with half the speed with which it hits the floor, find the height to which it rises after the rebound.

19. A ball is dropped from a height h on to a stone floor. If it rebounds with half the speed with which it hits the floor, find the height to which it rises after the rebound.

20. A ball is dropped from a height h on to a floor and rebounds to a height $\frac{1}{4}h$. Express the speed of rebound as a fraction of the speed with which the ball hits the floor.

CHAPTER 4

Force, The Laws of Motion

Elementary dynamics is based on three laws, first stated by Isaac Newton and published in his *Principia Mathematica* in 1687. These are experimental laws: they cannot be proved theoretically and we accept them only because results suggested by them agree with observed results. Recent observations have shown that in certain unusual situations, such as bodies travelling near the speed of light, these laws need modification. But they are quite adequate as a basis for our studies when the mass of a body is constant.

Newton's First Law

The weight of a man gave us our first example of a force. Others are the tension in a string, the thrust in a metal girder, the push we give a toboggan or lawn-mower, and the air-resistance to a cyclist.

Newton's First Law of Motion states that a body continues in its state of rest or of uniform motion in a straight line unless it is acted on by an external force.

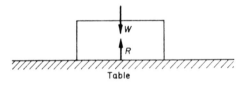

Fig. 11

Consider a book lying on a table. It is not moving and hence is said to be in equilibrium. Are there any forces acting on it? Certainly there is at least one, the weight W of the book. There is also another, the reaction R between the book and the table. These two forces are equal and opposite and cancel each other.

There is said to be no resultant force acting and the book is in statical equilibrium.

It is perhaps appropriate to state here that Statics is the study of bodies at rest whereas Dynamics is the study of bodies in motion.

Now consider a train which has accelerated from a station. There comes a time when the speed of the train remains constant. What forces are then acting on the train? These are the tractive or pulling force of the train and the resistances to motion. Since the speed of the train is constant, these two forces are equal and opposite. The train is not in equilibrium but is running at constant speed and will continue to do so unless there is some change in the forces acting. We might say that the train is in dynamical equilibrium. Of course, to give the train its velocity, the tractive force must at some earlier stage have been greater than the resistances against motion, but immediately the two forces become equal the train continues at constant speed.

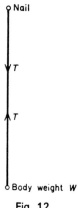

Fig. 12

Consider a string tied to a nail and supporting a body weight W at its other end. Since the body is in equilibrium, there must be an upward force equal to W acting on it. This is the tension in the string. The tension in a string always pulls inward at both ends; the tension at one end is equal and opposite to the tension at the other end. So the string pulls downward on the nail with a force T which is also equal to W. The nail is supported by the upward force equal to W due to the wall into which it is nailed.

If a piece of light wire with a body weight W attached to one end is held firmly in the hand at the other end with the body vertically above the hand, the body is supported by the thrust in the wire. This thrust must be equal and opposite to the weight W. A thrust acts outward at each end of a wire under strain and the

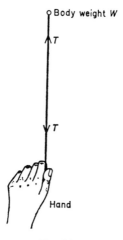

Fig. 13

thrust at one end is equal and opposite to the thrust at the other. A wire may take the place of a string, i.e. a wire may be in tension; but a string cannot take the place of a wire in thrust as it will crumple. The thrust at the other end of the wire is balanced by the force between the hand and the wire.

Next consider a book on a table with a gradually increasing horizontal force F pushing it. There are at least three forces acting on the book; its weight, the reaction between book and table and the force F. The weight and the reaction of the table balance each other as we have already seen. We know from experience that if we push the book with a small force, it does not move.

This means that there must be another force acting which we call the friction force between book and table. This adjusts itself to balance F provided it is not called upon to get greater than a certain maximum. When F becomes too large, the friction force cannot increase any more to balance F and the book moves. This will be considered later in the chapter on friction. In any problem it is essential that a diagram should be drawn and the forces acting on a body clearly marked. Remember that action at a distance is a rare occurrence. It does occur with gravity and with magnetic attractions but generally a force on a body is produced by virtue of the body being in contact with something which

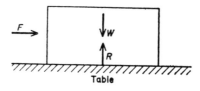

Fig. 14

produces a force. If you hold a string in your hand and the string supports a weight W hanging vertically, your hand does not produce a force acting on the book. Your hand keeps the string taut and it is the tension in the string which supports the weight.

Acceleration produced by a force: Newton's Second Law

If a body has a force acting continuously on it, the force produces an acceleration in the body in the direction of the force.

There are two things about which we must be careful when we think of this statement. First the force must be continuous. If we give a stone on ice a sharp blow, an acceleration is produced for the very short time contact exists. In this short time the stone acquires a velocity and then the acceleration ceases. The stone will continue to move indefinitely with this velocity unless other forces such as friction are brought into play. A short, sharp blow such as this is called an impulse and its apparent effect is to give the body a constant velocity. Secondly, by the force acting on a body we mean the resultant force. If two forces are acting on a body and these two forces are equal and opposite, an acceleration will not be produced. Newton's Second Law of Motion tells us that the acceleration produced in a body is proportional to the force acting. If the force on a body is doubled, the acceleration is doubled; if the force is trebled, the acceleration is trebled and so on. Again we must be careful to realise that force here means resultant force. Suppose that a force F produces an acceleration a in a body of mass m. A force of $2F$ will produce an acceleration $2a$ if and only if there is no other force acting in the direction of F.

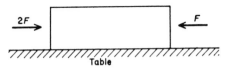

Fig. 15

Suppose as shown in Fig. 15, that a book lying on a smooth table has a horizontal force $2F$ acting in one direction and a force F acting in the opposite direction. There is a resultant force $(2F - F)$ or F acting on the book and so an acceleration a, say, is produced. If the forward force on the book is increased to $3F$ while the force F remains unaltered, the resultant force acting on the book is $2F$. Since the resultant force is doubled, the acceleration is doubled and is now $2a$. So a force $2F$ produces an acceleration a while a force $3F$ produces an acceleration $2a$, and acceleration is proportional to the resultant force.

Again the acceleration obviously depends on the mass of the body which is being accelerated. It is reasonable to assume that if we double the mass of the body, it will require double the force to produce the same acceleration; in other words, the force is directly proportional to the mass moved as well as to the acceleration.

If a resultant force F produces an acceleration a in a body of mass m, then a force $2F$ will produce an acceleration a in a body of mass $2m$.

The newton

The unit of force is the newton which is defined as that force which will produce an acceleration of 1 m s^{-2} in a mass of 1 kg.

1 newton (abbreviated to 1 N) produces an acceleration of 1 m s^{-2} in a mass of 1 kg; therefore a force of g newtons produces an acceleration of g m s^{-2} in a mass of 1 kg. But we know that a body of mass 1 kg falls freely under gravity with an acceleration of g m s^{-2}. So when a body of mass 1 kg is falling freely under gravity there must be a force of g newtons acting on it. Hence a mass of 1 kg has a weight of g newtons.

It follows from the definition that:

1 N produces an acceleration of 1 m s^{-2} in a mass of 1 kg;
2 N produce an acceleration of 2 m s^{-2} in a mass of 1 kg;
6 N produce an acceleration of 2 m s^{-2} in a mass of 3 kg.

Notice that the force in newtons is the product of the mass in kg and the acceleration in m s^{-2}.

So for a force of F newtons acting on a body of mass m kg, the acceleration produced, a m s^{-2}, is given by

$$F = ma.$$

This is Newton's Second Law, one of the most important equations in mechanics.

Example 1. *What force will produce an acceleration of 0.5 m s^{-2} in a car of mass 750 kg?*

$$F = ma = 750(0.5) = 375.$$

The force required is 375 N.

Example 2. *Find the retardation produced by frictional resistances of 45 N acting on a skater of mass 75 kg.*

$$F = ma$$
$$\Rightarrow 45 = 75a$$
$$a = \tfrac{45}{75} = \tfrac{3}{5}.$$

The retardation is 0.6 m s^{-2}.

Example 3. *A car of mass 700 kg accelerates from 3 m s^{-1} to 8 m s^{-1} in 2 s. Find the accelerating force.*

First find the acceleration.

$$v = u + at$$
$$\Rightarrow 8 = 3 + 2a$$
$$a = 2\tfrac{1}{2}.$$

Then
$$F = ma$$
$$\Rightarrow F = 700(2\tfrac{1}{2})$$
$$F = 1750.$$

The accelerating force is 1750 N.

Example 4. *A mass of 50 kg is moved vertically by a rope. Find the tension in the rope when:*

(i)   *the mass moves with a constant velocity of 2 m s^{-1},*
(ii) *the mass accelerates upwards at 4 m s^{-2},*
(iii) *the mass accelerates downwards at 4 m s^{-2}.*

The weight of the mass is 50 g newtons.

Fig. 16

There are two forces acting on the mass as shown in Fig. 16; the weight and the tension in the rope.

(i) As the acceleration is zero, the resultant force is zero.

∴ $T = 50g.$

The tension in the rope is 50g newtons or 490 N.

(ii) Let the tension in the rope be T newtons. The resultant upward force is $(T - 50g)$ newtons.

$$F = ma$$
$$\Rightarrow T - 50g = 50(4)$$
$$\text{i.e.,} \quad T = 490 + 200$$
$$\text{i.e.,} \quad T = 690.$$

The tension is 690 N.

(iii) The mass is accelerating downwards. The resultant force downwards is $(50g - T)$ newtons.

$$F = ma$$
$$\Rightarrow 50g - T = 50(4)$$
$$\text{i.e.,} \quad T = 490 - 200$$
$$\text{i.e.,} \quad T = 290.$$

The tension is 290 N.

EXERCISE 6

1. Find the acceleration produced if a ball of mass 0.7 kg is acted on by a force of 35 N.
2. If a force of 10 N acts on a mass of 2 kg, what acceleration is produced?
3. Calculate the force which produces an acceleration of 6 m s^{-2} in a car of mass 800 kg.
4. The mass of a balloon is 1500 kg. Assuming the gas in the balloon produces a constant lifting force of 20 000 N, find the acceleration of the balloon.
5. What force is necessary to give a train of mass 2×10^5 kg an acceleration of 0.5 m s^{-2}?
6. A toy truck of mass 5 kg has a forward force of 10 N acting on it. If the resistance to motion is 5 N, find the acceleration of the truck.
7. A car of mass 700 kg moving at 15 m s^{-1} is brought to rest in 5 m. Find the braking force.
8. A car of mass 700 kg moving at 15 m s^{-1} is brought to rest in 2 s. Find the braking force.
9. A horse pulls a truck of mass 500 kg along a level track with a force of 500 N. If the resistance to motion is 300 N, find the acceleration of the truck.

10. A small plane of mass 10 000 kg is flying at 70 m s^{-1}. The air resistance is 5000 N. If the thrust of the engines is 20 000 N, find the acceleration.

11. A shell of mass 15 kg has a thrust of 3×10^5 N acting on it. Find the acceleration of the shell.

12. A stone of mass 2 kg falls through water with an acceleration of 3 m s^{-2}. Find the resistance of the water.

13. The tension in the rope raising a bucket from a well is 30 N. If the acceleration of the bucket is 3 m s^{-2}, find the mass of the bucket.

14. A force of 120 N acts horizontally for 15 s on a truck of mass 1000 kg on horizontal rails. Find the velocity of the truck at the end of the 15 s.

15. A mass of 40 kg slides down a plane with an acceleration of 5 m s^{-2}. Find the accelerating force on the mass.

16. A trolley of mass 40 kg sliding down a plane acquires a speed of 10 m s^{-1} 2 s after starting from rest. Find the accelerating force on the trolley.

17. A bullet of mass 16 g travelling at 150 m s^{-1} embeds itself into a piece of wood. If it penetrates 20 cm into the wood, find the resistance of the wood to the motion of the bullet.

18. A speed boat of mass 5000 kg has an acceleration of 4 m s^{-2}. If the resistance of the water is 30 000 N, find the thrust of the engines.

19. A mass of m kg is acted on by a force of F N. If the resistance to motion is R N, find the acceleration of the mass in m s^{-2}.

20. A mass of m kg has an acceleration of a m s^{-2}. If the resistance to motion is R N, find the force acting in newtons.

Newton's Third Law

Newton's Third Law states that action and reaction are equal and opposite. As an illustration of this we consider a box of mass 4 kg lying on a table. The weight of the box is $4g$ N and this force is balanced by the upward reaction of the table on the box which is therefore equal to $4g$ N. Newton's law tells us that the box presses down on the table with a force of $4g$ N and so the forces acting on the table are the reaction with the box, $4g$ N, its own weight, say, $12g$ N and the reactions of the floor on the legs of the table. The sum of these reactions must therefore be $(4g + 12g)$ N or $16g$ N.

In Fig. 17, $R_1 + R_2 + R_3 + R_4 = 16g$ N.

Notice that we could have considered the box and table as one entity. Then the forces acting on the table are the reactions of the

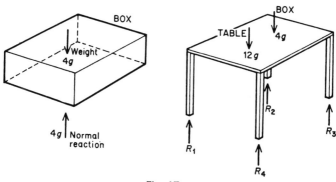

Fig. 17

floor on the legs, the weight of the box and the weight of the table. The reaction forces at the legs must therefore in total equal the sum of the two weights of the table and box, i.e. 16 N. Here we do not bring in the reaction between box and table because for the composite body it is an internal force. It is similar to the forces which act at the joints between the table legs and the table top.

As a second illustration, we consider a car pulling a caravan.

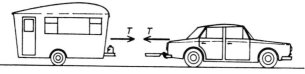

Fig. 18

The force which pulls the caravan along is the tension in the tow bar and not the thrust of the engine of the car.

The horizontal forces on the car are the thrust of the engine forwards and the tension in the tow bar backwards (neglecting resisting forces). The effect of the caravan is to reduce the acceleration of the car because of the tension in the tow bar. When the velocity of the car becomes constant, the tension in the tow bar is just sufficient to overcome any resistance on the caravan; the thrust of the engine is just sufficient to overcome the resistance on the car and the resistance on the caravan.

The acceleration of the car may be calculated by considering the car and caravan together and the resultant forward force as acting on the total mass.

Some applications of Newton's Third Law may be puzzling. What happens in a tug of war? Team A must pull team B with exactly the same force as team B pulls team A and so one might imagine that every tug of war must be a draw. To find the fallacy in this argument, it is necessary to consider more closely the forces acting. It is true that the tension, T, in the rope pulls team

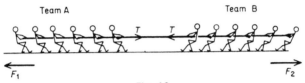

Team A   Team B

Fig. 19

A just as hard as it pulls team B, but what other horizontal forces are acting? They are the frictional forces between the feet and the ground, F_1 and F_2. If F_1 is larger than F_2, team A wins; if F_2 is larger than F_1, then B wins. The tension will always have a value between F_1 and F_2. It cannot be larger than both, because then both teams would move in and the rope would become slack; it cannot be smaller than both, because then both teams would move out and the rope would either slip or break.

Example 1. *A box of mass 4 kg on the tail board of a lorry does not slip when the lorry accelerates at 1 m s^{-2}. Find the friction force acting on the box.*

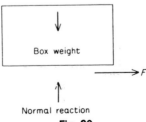

Fig. 20

The only horizontal force acting on the box is the friction force between box and lorry which must therefore act forward on the box and backwards on the lorry. Suppose this force is F N. Since the box does not slip, the acceleration of the box is the same as that of the lorry, namely 1 m s^{-2}. So the force F N produces in the mass of 4 kg an acceleration of 1 m s^{-2}.

$$F = ma = 4(1).$$

The friction force is 4 N.

Motion of connected bodies

The following examples investigate the motion of connected bodies.

Example 2. *A taut string passes over a smooth pulley at the edge of a smooth table. It has a mass of 5 kg attached to one end and hanging freely and a mass of 3 kg attached to the other end which lies on the table. Find the tension in the string and the acceleration of either mass.*

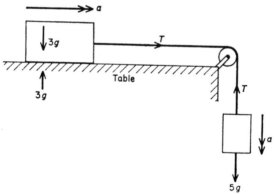

Fig. 21

Since the pulley is smooth the tension upward on the 5 kg is equal to the horizontal pull on the 3 kg. When a string passes over a smooth peg or a smooth pulley which rotates freely without friction, the tensions in the two parts of the string are equal. When a string passes over a rough pulley, the tensions in the string on the two sides of the pulley will be different due to the fact that the pulley does not turn effortlessly. The smooth pulley is an ideal which is never met in real life. Most questions on strings also assume that the string is inextensible; in practice, there is always some extension, however minute.

Suppose the tension in the string is T N and the common acceleration of the masses is a m s^{-2}.

The vertical forces on the 3 kg mass balance each other; the only horizontal force is T which produces in the mass an acceleration of a m s^{-2}. From $F = ma$ we have

$$T = 3a. \tag{i}$$

The only forces acting on the 5 kg mass are the tension T and the weight $5g$. The mass moves downwards with acceleration a and the resultant downward force is $(5g - T)$ N. Using $F = ma$,

$$5g - T = 5a. \tag{ii}$$

Adding (i) to (ii);

$$5g = 8a.$$

The acceleration is $\frac{5}{8}g$ m s^{-2}.
Substituting in $T = 3a$, we have $T = \frac{15}{8}g$. The tension is $\frac{15}{8}g$ N.

Example 3. *A string passes over a smooth pulley with masses of 3 kg and 5 kg attached to the ends of the string and hanging vertically. Find the acceleration of either mass and the tension in the string.*

Suppose the tension in the string, which is the same throughout, is T N. The 5 kg mass will obviously fall and suppose its acceleration downwards is a m s^{-2}. The acceleration of the 3 kg mass is then a m s^{-2} upwards.

The forces acting on the 5 kg mass are $5g$ N down and T N up. The resultant downward force is $(5g - T)$ N. Using $F = ma$,

$$5g - T = 5a. \tag{i}$$

The forces acting on the 3 kg mass are $3g$ N down and T N up.

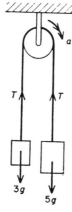

Fig. 22

The resultant upward force is $(T - 3g)$ N. Using $F = ma$,

$$T - 3g = 3a. \qquad \text{(ii)}$$

Adding (i) to (ii);

$$2g = 8a,$$

$$a = \tfrac{1}{4}g.$$

The acceleration is $\tfrac{1}{4}g$ m s^{-2}.

Substituting in $T - 3g = 3a$, we have $T = 3g + \tfrac{3}{4}g$. The tension in the string is $\tfrac{15}{4}g$ newtons.

If we had assumed that the direction of the acceleration was reversed, i.e. that the 5 kg mass was moving upwards, the equations would have compensated for the mistake and we should have found that the 5 kg mass was moving upwards with an acceleration of $-\tfrac{1}{4}g$ m s^{-2}.

EXERCISE 7

1. A box of mass 8 kg on the tail board of a lorry does not slip when the lorry accelerates at $\tfrac{1}{2}$ m s^{-2}. Find the friction force acting on the box.

2. A car of mass 1000 kg tows a caravan of mass 600 kg. If the driving force of the engine is 240 N, find the acceleration of the car and the tension in the tow bar.

3. A man of mass 80 kg is descending in a lift which has a downward acceleration of 2 m s^{-2}. Find the force between the man and the floor.

4. A train of mass 4×10^4 kg pulls a coach of mass 3×10^4 kg. Supposing that there is a resistance of 1 newton per 100 kg acting on both coach and train and that the driving force of the train is 3500 N, find the acceleration of the train. Find also the tension in the coupling.

5. A balloon of mass 300 kg is drifting horizontally when 30 kg of ballast is thrown out. Find the acceleration of the balloon.

6. A man in a lift holds a spring balance with a mass of 3 kg hanging from it. The lift descends with a constant acceleration of 1.5 m s^{-2}. What is the reading of the spring balance if it is calibrated in newtons?

7. A bucket full of water is being pulled from a well and has a total mass of 15 kg. If the bucket has an acceleration of 0.6 m s^{-2}, find the tension in the rope.

8. A mass of 9 kg lies on a smooth table and is connected by a string, which passes over a smooth pulley at the edge of the table, to a mass of 5 kg hanging vertically. Find the acceleration of the masses.

9. In question 8, supposing that a friction force of 14 N acts on the 9 kg mass, find the acceleration of the masses.

10. Two masses each of 7 kg are connected by a taut string which passes over a smooth pulley at the edge of a table. One mass lies on the table and the other hangs vertically. Find how far the hanging mass descends in 2 s.

11. Two masses of 5 kg and 9 kg are connected by a taut string which passes over a smooth fixed pulley so that both masses hang freely. Find the acceleration of either mass.

12. Two scale pans each of 6 kg are connected by a taut string which passes over a smooth fixed pulley. A mass of 2 kg is placed in one pan and motion takes place. Find the acceleration of either scale pan and the reaction between pan and mass.

13. A mass of 10 kg lies on a smooth table at a distance of 7 m from its edge and is connected by a taut string passing over the end with a mass of 4 kg hanging freely. How long does it take the 10 kg mass to reach the edge of the table?

14. Two buckets each of mass 5 kg are suspended from the ends of a taut rope passing over a smooth pulley. A stone of mass 4 kg is dropped into one bucket so that the bucket and stone move downwards with a velocity of 3.5 m s^{-1}. Find
   (i) the acceleration of the buckets,
   (ii) the distance moved by either bucket in 2 s.

15. Masses of 6 kg and 4 kg are fixed to the ends of a taut string which hangs over pulleys at the edges of a smooth table as shown in Fig. 23. A mass of 4 kg attached to the string rests on the table. Find the acceleration of the 6 kg mass.

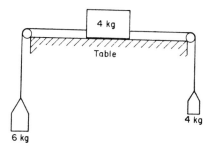

Fig. 23

16. A taut string passes over a rough pulley and has masses of 10 kg and 4 kg attached to its ends. The part of the string on the side of the 10 kg mass has a tension which is 10 N greater than the tension in the other part of the string. Find the acceleration of either mass.

17. A mass of 5 kg is raised vertically by a force of 100 N. Find its acceleration.

18. I buy $\frac{1}{2}$ kg of sweets in a store and go down in the lift with the box hanging from my finger. The lift has a downward acceleration of 1 m s^{-2} at the beginning and a retardation of 2 m s^{-2} at the end. How heavy in newtons does the parcel seem at the beginning and end of the lift's descent?

19. A mass of 2 kg is pulled along smooth horizontal rails directly towards a smooth pulley by a string which passes over the pulley and is attached to a mass of 1.5 kg hanging vertically. Find the acceleration of either mass.

20. A pinnace of mass 15 000 kg takes in tow a launch of mass 10 000 kg and the resistance of the water is 1 N per kg. Find the tension in the hawser when the pinnace is accelerating at 0.5 m s^{-2}.

21. Two masses of m kg and M kg are connected by a string passing over a smooth pulley at the edge of a smooth table. The mass m lies on the table and the mass M hangs vertically. Find the acceleration of either mass.

22. Two masses of m kg and M kg are connected by a string passing over a smooth pulley so that each mass hangs vertically. Find the acceleration with which M descends, if M is greater than m.

23. Fig. 24 shows an arrangement of smooth, weightless pulleys used to lift a mass of 4 kg. Find, in newtons, the tension T
  (i) if the 4 kg mass is in equilibrium,
  (ii) if the 4 kg mass has an upward acceleration of 2 m s^{-2}.

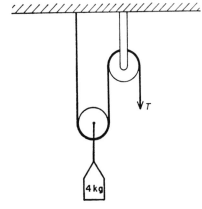

Fig. 24

24. Masses of 1 kg and 0.96 kg are connected by a string passing over a smooth pulley. After 2 s, the velocity of each mass is 0.38 m s^{-1}. From these figures, find an approximate value for g in m s^{-2}.

25. A mass of 5 kg on a horizontal table is set in motion by a mass at the end of a string passing over a pulley at the edge of the table. When a mass of 50 g is fixed to the free end of the string, the 5 kg mass moves with constant velocity. What will be the acceleration when a mass of 100 g is fixed to the free end instead?

CHAPTER 5

Moments

Moment of a force

We have so far considered particles, or bodies which may in the context be treated as particles. A particle is a fictitious element without size. We now go on to consider the effect of forces on bodies which have size as well as weight. Consider first a uniform bar AB as in Fig. 25 which is considered to have length but not any other dimension.

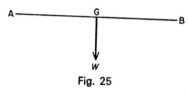

Fig. 25

Each of the elements comprising the bar has a weight which acts vertically downwards. The number of these elements is infinite but their weights are equivalent to a single force, the total weight W, acting at a point in the bar: this point is called the centre of gravity (see Chapter 10). In the case of a uniform bar the centre of gravity is, by symmetry, at the mid point of the bar and the total weight of the bar may be considered as acting at its mid point. Similarly, the centre of gravity of a uniform rectangular lamina is at the intersection of its diagonals; the centre of gravity of a uniform circular lamina is at its centre.

Suppose the bar is supported by a knife edge at G and has a weight of 10 N. The reaction at the knife edge is also 10 N and the bar is in equilibrium, a position very difficult to hold in practice but one theoretically possible. Now suppose a weight of 10 N is suspended from B. What will happen? The centre of gravity cannot move vertically and so the reaction at the knife edge increases to 20 N to prevent vertical motion. But from our practical experience we know that the rod will turn due to the force acting at B.

Another good example is provided by an open door. Let one person try to shut the door by pushing with his little finger at the edge remote from the hinge and another try to keep the door open by pushing with his whole hand very near the hinge. The little finger is capable of shutting the door in spite of the greater force acting near the hinge. So the nearer the force is to the hinge, the less its turning effect; the further away the force is from the hinge, the greater its turning effect.

If a body is in equilibrium, there must not be any tendency for the body to move or for the body to rotate. If the forces are all parallel, the body obviously can move only in the direction of the forces themselves. For equilibrium, therefore, the algebraic sum of the parallel forces must be zero, which means that the sum of the forces in one direction must equal the sum of the forces in the opposite direction. This condition is not sufficient, however, to prevent rotation. The turning effect of a force about a point is called its moment and is measured by the product of the force and the perpendicular distance from the point to the force. The turning effect of 1 N about a point from which its perpendicular distance is 1 m is 1 N m; the moment of a force of 4 N about a point distant 3 m from it is 12 N m.

It should be pointed out that every body turns about an axis and not a point. Moments should therefore always be taken about an axis and not a point. Many of our examples, however, are concerned with forces which all act in one plane. It is usual then to speak of taking moments about a point when in fact we mean about the axis through the point perpendicular to the plane containing the forces. The moment of a force about a point may be either counter-clockwise (by convention this is called positive) or clockwise (called negative). In particular, if the line of action of the force passes through the point about which we are taking moments, its moment is zero.

Bodies in equilibrium

If a body is in equilibrium, the algebraic sum of the moments of the forces *about any point* must be zero. The weight of the body, itself, must be considered as one of the forces acting and the weight of the body may always be considered as a single force acting at its centre of gravity.

The sum of the moments of the forces must be zero about whatever point we choose. The choice of the point about which we take moments is only important in that, if a suitable point is chosen, the working will be simplified. The point chosen is usually a point through which an unknown force acts so that this force does not appear in the resulting equation.

Example 1. *A uniform bar is 4 m long and of weight 20 N. Weights of 30 N and 40 N are suspended from its ends. Where must the bar be supported so that it rests horizontally?*

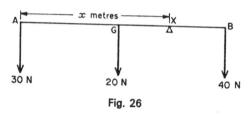

Fig. 26

Let AB be the bar and G its centre of gravity. Suppose the bar must be supported at X and that the distance AX = x m.

First method. Take moments about X.
The moment of the 30 N is $30(x)$ N m counter-clockwise.
The moment of the 20 N is $20(x - 2)$ N m counter-clockwise.
The moment of the 40 N is $40(4 - x)$ N m clockwise.
The moment of the force at X is zero.
Therefore

$$30x + 20(x - 2) = 40(4 - x)$$
$$50x - 40 = 160 - 40x$$
$$90x = 200$$
$$x = 2\tfrac{2}{9}.$$

Second method. As the bar is in equilibrium, the upward force of the support at X must be equal to $(30 + 20 + 40)$ N or 90 N. Take moments about A.

The moment of the 30 N is zero.
The moment of the 20 N is 20(2) N m clockwise.
The moment of the 90 N is 90(x) N m counter-clockwise.
The moment of the 40 N is 40(4) N m clockwise.
Therefore

$$20(2) + 40(4) = 90x$$
$$40 + 160 = 90x$$
$$x = 2\tfrac{2}{9}.$$

The point of support is $2\tfrac{2}{9}$ m from A or approximately 2.2 m.

Example 2. *A bar AB of length 1 m and mass 8 kg and with its centre of gravity 40 cm from A rests on supports at A and B. Masses of 2 kg, 3 kg, and 4 kg are suspended at points 20 cm, 60 cm, and 80 cm from A. Find the reactions at the supports.*

Fig. 27

The weights in newtons are 2g, 8g, 3g, and 4g. The distances from A to their lines of action, in metres, are 0.2, 0.4, 0.6 and 0.8 respectively.

Let the reactions at A and B be P N and Q N respectively. Taking moments about A,

    the moment of P is zero,
    the moment of 2g is 2g(0.2) N m clockwise,
    the moment of 8g is 8g(0.4) N m clockwise,
    the moment of 3g is 3g(0.6) N m clockwise,
    the moment of 4g is 4g(0.8) N m clockwise,
and the moment of Q is Q(1) N m counter-clockwise.

$$\therefore \qquad Q = 2g(0.2) + 8g(0.4) + 3g(0.6) + 4g(0.8)$$
$$= 0.4g + 3.2g + 1.8g + 3.2g$$
$$= 8.6g.$$

But $P + Q = 17g$, Hence

$$P + 8.6g = 17g$$
$$P = 8.4g.$$

The reactions at the supports are $8.4g$ N and $8.6g$ N.

EXERCISE 8

1. A light rod of length 4 m carries masses of 6 kg and 10 kg at its ends A and B respectively. The rod with the masses will balance about a point P. Find the distance AP.

2. A uniform rod of length 4 m and mass 4 kg carries masses of 6 kg and 10 kg at its ends A and B respectively. It balances about a point L. Find the distance AL.

3. A non-uniform horizontal rod **AB** of length 2 m is supported at its two ends. The reaction at A is $4g$ N and the reaction at B is $6g$ N. Find the mass of the rod and the distance of the centre of gravity of the rod from A.

4. A light horizontal beam rests freely with its ends on two supports 2 m apart. This beam carries a load of 20 N at 0.8 m from the left-hand support. Calculate the reaction at the supports.

5. A uniform horizontal beam of mass 4 kg and length 2 m rests upon supports at its ends. It has a mass of 2 kg attached 0.8 m from one end. Find the reaction at the supports.

6. A uniform beam of mass 3 kg and length 4 m is supported at its midpoint. It has a mass of 2 kg attached at one end. What mass must be attached at a distance of 1 m from the midpoint to keep the bar horizontal?

7. A load is placed on a uniform heavy plank of mass 10 kg and length 4 m, which is supported at its two ends. Find the position of the load if the reactions at the supports are $8g$ N and $12g$ N.

8. A uniform horizontal bar of length 6 m and mass 20 kg is supported at its ends. Find the reactions at the supports when a mass of 8 kg is suspended 1 m from one end and a mass of 10 kg is suspended 2 m from the other end.

9. A horizontal beam of length 6 m is supported at its two ends. If the reactions at the ends are 100 N and 200 N, find the position of the centre of gravity of the bar.

10. A boy and a man carry a uniform horizontal ladder 8 m long and of mass 50 kg. The boy supports one end of the ladder with a force of $20g$ N. How far from its centre does the man support the ladder?

11. A uniform bar of mass 2 kg carries loads as shown in Fig. 28. What are the readings in newtons of the two spring balances which support the bar?

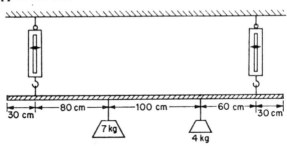

Fig. 28

12. A horizontal bar 2 m long mass 2 kg is supported at its two ends. A mass of 7 kg is hung from the bar 50 cm from one end and a mass of 5 kg is hung 30 cm from the other end. Find the reactions at the supports.

13. A uniform horizontal rod is 1 m long and has at one end an extension of length 20 cm of the same cross section as the rod but of different material. If the rod balances about a point 80 cm from the non-extended end, find the ratio of the densities of the two materials.

14. The metal part of a telescope is made up of three tubes each 16 cm long. The masses of the tubes are 200, 300 and 500 g and the weight of each tube acts at its mid-point. Find the distance from the end of the lightest tube where a support should be placed to keep the telescope horizontal, when the telescope is open.

15. A uniform trestle table is 4 m long and mass 30 kg. The trestles are placed 80 cm from each end and masses of 4 kg and 10 kg are placed on the table at distances of 40 cm and 120 cm from one end. Find the load supported by each trestle.

16. The handles of a wheelbarrow are 120 cm from the wheel axle and 40 cm measured horizontally from the legs. A vertical force of $10g$ N applied at the end of the handles will just lift the legs off the ground. A vertical force of $25g$ N applied at the axle will just raise the wheel off the ground. Find the total mass of the wheelbarrow.

17. A pupil being taught to swim is held by a vertical rope attached to one end of a horizontal, uniform bar of mass 2 kg and length 2 m. The other end of the bar is held by the instructor by one hand and his other hand supports the bar so that the distance between his two hands is 40 cm. The tension required in the rope is $5g$ N. Find the force which each hand must exert.

18. A beam of length 2 m is held horizontally by two vertical strings attached to its ends. A mass of 7 kg is suspended from a point 60 cm from one end. The tension in the string at this end is $6.5g$ N and the tension in the other string is $3g$ N. Find the mass of the beam and the point where its weight acts.

19. A uniform plank 6 m long is held in a horizontal position by supports at its two ends. The mass of the plank is 20 kg. Two masses each of 30 kg are placed on the plank at distances of 1 m and 4.5 m from one end. Calculate the reactions at the supports.

20. From the ends of a horizontal uniform bar, of length 50 cm and mass 500 g, are hung two scale-pans each of mass 100 g. What mass must be placed in one scale-pan so that the bar will balance on a knife edge 20 cm from that pan?

Couples

A couple consists of two equal and opposite forces which are not collinear, as shown in Fig. 29. Since the forces are equal and opposite, there is no tendency for the body to move in any direction but the couple obviously has a turning effect. Examples of a

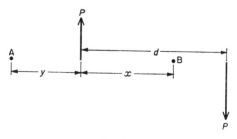

Fig. 29

couple are common and include unscrewing the lid of a jar, turning on a tap and turning the handle of a door. Another good example is the use of a turncock for turning off the main supply of water to a house. The moment of a couple is constant about any point in its plane and is equal to the product of one of the forces and the perpendicular distance between the lines of action of the forces. The moment of the couple illustrated in Fig. 29 is Pd clockwise. The only system of forces which will balance a couple is another couple or its equivalent. To show that a couple has the same turning effect about any point in the plane, consider the two points A and B as shown in Fig. 29.

The moment about A is $P(d + y) - Py$, i.e. Pd clockwise.

The moment about B is $Px + P(d - x)$, i.e. Pd clockwise.

Example. *ABCD is a rectangular plate with forces of 6 N acting along AD and CB. If AD = 2 m and AB = 3 m, find the forces necessary along CD and AB to keep the plate in equilibrium.*

Since the given forces form a couple, they need another couple to balance them. The couple given is clockwise so an anti-clockwise couple is required. Suppose the forces required are P N along CD and along AB.

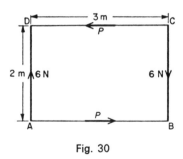

Fig. 30

The moment of the 6 N couple is 6(3) N m clockwise.
The moment of the P N couple is $P(2)$ N m anti-clockwise. Hence

$$2P = 18,$$
$$P = 9.$$

Parallel forces

Parallel forces are called *like* if they are in the same direction; *unlike*, if they are in opposite directions.

Parallel forces may always be replaced by a single force, parallel to the original forces, unless the forces form a couple.

Like forces

Suppose the two parallel forces P and Q act at A and B respectively Fig. 31(a). Their resultant Fig. 31(b) will be a force of $(P + Q)$ but where will it act? If it is equivalent to the two forces,

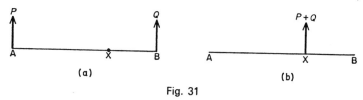

(a)   (b)

Fig. 31

it must have the same moment about any point we care to choose. The moment of the two original forces about A is $Q(AB)$. The moment of the resultant force about A is $(P + Q)AX$ where X is the point in which the line of action of the resultant cuts AB. Therefore

$$Q(AB) = (P + Q)AX$$

or

$$AX = \frac{Q}{P + Q} \cdot AB$$

Similarly

$$BX = \frac{P}{P + Q} \cdot AB$$

Hence

$$\frac{AX}{BX} = \frac{Q}{P}.$$

So the line of action cuts AB in the ratio $Q:P$.

Unlike forces

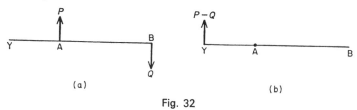

(a)   (b)

Fig. 32

Suppose as shown in Fig. 32(a) that two unlike forces P and Q act at A and B respectively and that P is greater than Q. The resultant Fig. 32(b) will then be $(P - Q)$ acting, say, at Y. By moments about A,

$$Q(AB) = (P - Q)YA$$

and Y must lie on BA produced to make both moments clockwise. Therefore

$$YA = \frac{Q}{P - Q} \cdot AB, \qquad YB = \frac{P}{P - Q} \cdot AB.$$

So Y is the point that divides AB externally in the ratio $Q : P$. The point Y is called the centre of the parallel forces and these applications may of course be extended to cases in which more than two forces are acting.

Non-parallel forces

If the forces are not parallel, the principle of moments still applies. The moment of the force P shown in Fig. 33 about O is $P(p)$ where p is the perpendicular distance from O to the line of action of P, produced if necessary.

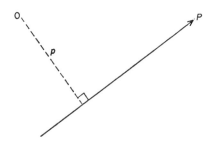

Fig. 33

Example 1. *A uniform bar AB is hinged at its mid-point C so that it can turn freely. A force of 10 N acts at A and the bar is kept horizontally in equilibrium by a string attached to B and making an angle of 30° with BA. Find the tension in the string.*

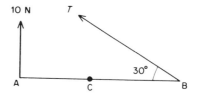

Fig. 34

The only forces acting are the force of 10 N, the tension and some force at the mid-point C. Take moments about C so that the unknown force and the weight of the bar have zero moment.

The moment of the 10 N force is $10l$ N m where l is half the length of the bar. The moment of T about C is $T(\frac{1}{2}l)$ N m. Hence

$$T(\tfrac{1}{2}l) = 10l,$$
$$T = 20.$$

The tension is 20 N.

Example 2. *One end of a string, 50 cm long, is fixed at a point O, and a bob of mass 2 kg is attached at the other end. The bob is kept in equilibrium by a horizontal force of F N so that it hangs 30 cm from the vertical through 0. Find the value of F*

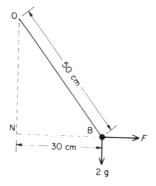

In Fig. 35, OB $= 50$ cm and NB $= 30$ cm. By Pythagoras' theorem, ON $= 40$ cm.

The only forces acting on the bob are its weight, the force F and the tension in the string. Take moments about O, so the moment of the tension is zero.

The moment of $2g$, (the weight in newtons), is $2g(0.3)$ N m.

The moment of F is $F(0.4)$ N m, Hence

$$4F = 6g,$$
$$F = 1.5g.$$

The force required is $1.5g$ N.

EXERCISE 9

1. Two like parallel forces of 4 N and 6 N are 2 m apart. How far is the line of action of the resultant force from that of the smaller force?

2. Two like parallel forces of 5 N and 7 N are 3 m apart. How far is the resultant from the smaller force?

3. Two unlike parallel forces of 8 N and 4 N are 2 m apart. How far is the resultant force from the smaller force?

4. Two unlike parallel forces of 8 N and 2 N are 3 m apart. How far is the resultant from the smaller force?

5. A rectangular plate ABCD has forces of 5 N acting along AB and CD. Given that AB $= 2.5$ m and that AD $= 3$ m, what forces along AD and CB will keep the plate in equilibrium?

6. Forces of 2 N act along each of the sides AB, BC, CA of an equilateral triangle ABC of side 1 m. Show that the moments of these forces about A, B and C are the same.

7. Forces F, F, $2F$ and $2F$ act along the sides $\overrightarrow{AB}$, $\overrightarrow{BC}$, $\overrightarrow{DC}$, $\overrightarrow{AD}$ of a square lamina ABCD. Show the lamina can be kept in equilibrium by pinning it to a fixed surface with just one pin, if the pin is inserted in a certain point.

8. A rectangular table 1 m by 1.5 m and mass 10 kg is supported by legs at the midpoint of each side. Find the least mass which, when placed on the table, will overturn it.

9. A circular table has three legs, placed at points on the circumference of the circle which are vertices of an equilateral triangle. If the mass of the table is 20 kg, find the least force which will overturn the table.

10. Forces, each of 4 N, act along the sides AB, BC, CD, DA of a square ABCD of side 2 m. Prove that the forces are equivalent to a couple and find the moment of the couple.

11. The drum of a windlass is 10 cm in diameter and the effort is applied to a handle 50 cm from the axis. What is the force necessary to raise a mass of 50 kg?

12. A light uniform bar is pivoted at its mid-point. A force of 30 N is applied vertically downwards at one end. What force at the other end making an angle of 30° with the bar is necessary to keep the bar horizontal?

13. A wheel of radius 30 cm and of mass 30 kg is being rolled up a step 6 cm high as shown in Fig. 36. Find the least horizontal

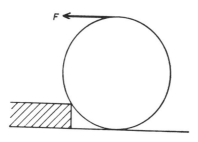

Fig. 36

force F which must be applied at the highest point of the wheel to turn it, assuming that it does not slip.

14. The bob of a pendulum of mass 2 kg is drawn aside so that the string attached to the bob makes an angle of 60° with the vertical. What horizontal force is necessary to keep the bob in equilibrium?

15. A nut on a machine is tightened by applying a force of 50 N at the end of a spanner. If the distance from the point of application of the force to the centre of the nut is 12 cm, find the couple acting on the nut and explain how it is obtained.

16. A mass of 50 kg is hung from one end of a rope 2 m long, the other end of which is fixed. Find the *least* force necessary to hold the mass aside from the vertical a distance of 1 m measured horizontally. In what direction is the force applied?

17. Forces of 5 N, 5 N, 5 N, 10 N act along the sides AB, BC, CD, DA respectively of a square ABCD of side 2 m. Find the resultant of these forces and the distance of its line of action from A.

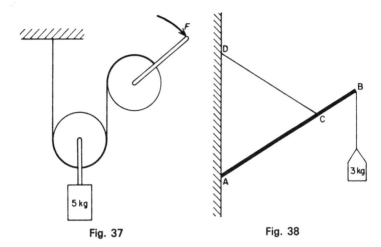

Fig. 37   Fig. 38

18. The mass of 5 kg on a grandfather clock is raised by turning a key with a lever 5 cm long. This turns a drum of radius 3 cm round which the cord passes as shown in Fig. 37. Find the force F, applied to the end of the lever, necessary to raise the mass.

19. A light bar AB of length 40 cm has a mass of 3 kg hanging from B. The bar is hinged at A and is supported by the chain CD as shown in Fig. 38. Find the tension in the chain given that AD is vertical and ACD an equilateral triangle of side 30 cm.

20. ABCD is a square of side 2 m. Forces of 4 N, 2 N and 4 N act along the sides DC, CB and BA respectively. Find the distance from A of the line of action of the resultant of the forces.

21. A metal sphere, mass 5 kg, rests against a smooth vertical wall. One end of a string, equal in length to the radius of the sphere, is attached to the sphere and the other end to the wall. Show by taking moments about the centre of the sphere that the line of the string, when produced, passes through the centre of the sphere. Find the tension in the string.

22. A nut is being tightened by a finger and thumb each applying a force of 20 N. If the distance between finger and thumb is 2 cm, find the moment exerted.

23 A capstan is being turned by four men pushing at the ends of spokes each 150 cm long. If each man exerts a force of 100 N, find the moment of the couple exerted on the capstan.

24. A table top lets down from a wall and is supported in a horizontal position by a light stay XY as shown in Fig. 39. If the centre of gravity of the table top is 48 cm from A, AY = 30 cm and AX = 40 cm, find the thrust in the stay if the mass of the table top is 5 kg.

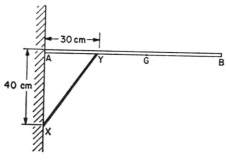

Fig. 39

25. The lid of a box 80 cm wide is uniform and of mass 2 kg. It is held open at an angle of 60° to the horizontal by a stay 20 cm long which makes equal angles with the horizontal and the lid of the box as shown in Fig. 40. Find the thrust in the stay.

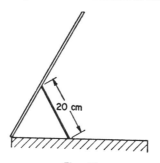

Fig. 40

CHAPTER 6

Vectors

Definition

A scalar quantity is one which requires its magnitude only to determine it whereas a **vector** needs more than one quantity to define it. Perhaps the most common vectors we meet are velocity and force which need both magnitude and direction to fix them.

A vector can be defined as a quantity which has both magnitude and direction and which can be added to other vectors by the law of addition set out in the following paragraphs.

The point (3,1) in a cartesian framework is obtained by moving 3 units along the x-axis and then 1 unit parallel to the y-axis. The line joining the origin to the point (3,1) and pointing towards that

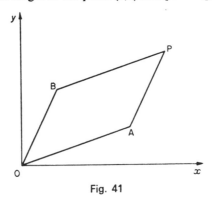

Fig. 41

point is called a *position vector*. If we add to this the position vector (1,2) we mean that we travel $(3 + 1)$ units along the x-axis and $(1 + 2)$ units parallel to the y-axis. We then arrive at the point (4,3). This is represented in Fig. 41 by the point P where P is found by completing the parallelogram OAPB. We can define a vector joining any two points in space as having a magnitude equal to the distance between the points and being in the direction of the line joining the two points. If the two points are B and P, then the vector is written $\overrightarrow{BP}$ and we see from Fig. 41 that this vector is equal and parallel to $\overrightarrow{OA}$.

To fix a point in space needs three cartesian coordinates and the coordinates of such a point may be written in the form $(1, -3, 4)$.

Addition of vectors

If $\overrightarrow{AB}$, $\overrightarrow{AC}$ are two vectors in space, then their sum is found by completing the parallelogram ABDC, the sum being represented by the diagonal $\overrightarrow{AD}$ as shown in Fig. 42(a).

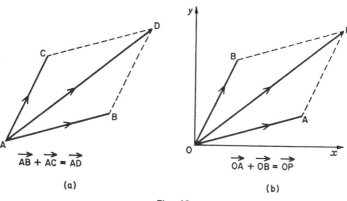

$$\overrightarrow{AB} + \overrightarrow{AC} = \overrightarrow{AD}$$

(a)

$$\overrightarrow{OA} + \overrightarrow{OB} = \overrightarrow{OP}$$

(b)

Fig. 42

If we were to add the two vectors $\overrightarrow{OA}$ and $\overrightarrow{OB}$ where O is the origin, then $\overrightarrow{OP} = \overrightarrow{OA} + \overrightarrow{OB}$ as shown in Fig. 42(b).

It is often convenient to denote $\overrightarrow{OA}$ by $\mathbf{a}$, $\mathbf{a}$ being the position vector of A, i.e. representing $\overrightarrow{OA}$. If $\overrightarrow{OB} = \mathbf{b}$, $\overrightarrow{OP} = \mathbf{p}$, then

$$\mathbf{p} = \mathbf{a} + \mathbf{b}.$$

This is the parallelogram law of addition.

Subtraction of vectors

If $\overrightarrow{OA}$ and $\overrightarrow{OB}$ are two vectors, $\overrightarrow{OA} - \overrightarrow{OB}$ is defined as $\overrightarrow{OA} + (-\overrightarrow{OB})$ and is equal to $\overrightarrow{OQ}$ in Fig. 43.

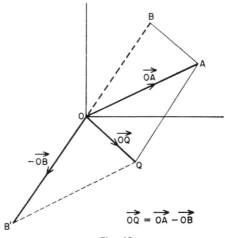

$$\overrightarrow{OQ} = \overrightarrow{OA} - \overrightarrow{OB}$$

Fig. 43

If $\overrightarrow{OQ} = \mathbf{q}$, then $\mathbf{q} = \mathbf{a} - \mathbf{b}$.

It is easily proved that OQAB is a parallelogram and therefore that OQ is equal and parallel to BA. Provided we are concerned with magnitude and direction only and not the actual position of the vector,

$$\overrightarrow{BA} = \overrightarrow{OA} - \overrightarrow{OB}.$$

Therefore the vector $\overrightarrow{BA}$ is represented by $\mathbf{a} - \mathbf{b}$.

Deduction

AB is equal and parallel to CD $\Leftrightarrow \mathbf{a} - \mathbf{b} = \mathbf{c} - \mathbf{d}$.

Multiplication of a vector by a scalar

If we add the vector $\mathbf{a}$ to the vector $\mathbf{b}$, we write the sum as $\mathbf{a} + \mathbf{b}$ It is logical to define $2\mathbf{a}$ as $\mathbf{a} + \mathbf{a}$, and to extend this to any scalar multiple of **a**. Thus

$$n\mathbf{a} = \mathbf{a} + \mathbf{a} + \mathbf{a} \ldots \ldots + \mathbf{a},$$

to n terms if n is an integer, and we define $(1/n)\mathbf{a}$ to be such a vector that

$$\mathbf{a} = \frac{1}{n}\mathbf{a} + \frac{1}{n}\mathbf{a} + \frac{1}{n}\mathbf{a} + \frac{1}{n}\mathbf{a} + \ldots \frac{1}{n}\mathbf{a},$$

to n terms.

The extension to a multiple of the form $(m/n)\mathbf{a}$ follows immediately. We need the scalar multiple of a vector in many situations, e.g. the distance travelled in a given time with given velocity. The multiplication of a vector by a vector will be useful later, but is outside the range of this book.

The unit vector

A unit vector in a given direction is a vector with unit magnitude in that direction.

If we have a vector **a**, its magnitude is often written $|\mathbf{a}|$; thus

$$\mathbf{a} = |\mathbf{a}| \, \hat{\mathbf{a}},$$

where $\hat{\mathbf{a}}$ is the unit vector in the direction of **a**.

We shall let **i** represent the unit vector along the axis of x and **j** the unit vector along the axis of y. Any point on the positive x-axis will have position vector $\lambda\mathbf{i}$ where λ is positive, and any point on the negative x-axis $-\lambda\mathbf{i}$. Similarly any point on the y-axis has position vector $\lambda\mathbf{j}$; if λ is positive, the point will be on the positive axis, if λ is negative, the point will be on the negative part of the y-axis.

For example, the position vector of the point $(3,0)$ is $3\mathbf{i}$; the position vector of the point $(0, -2)$ is $-2\mathbf{j}$.

Since the position of the point $(3,4)$ is found by moving 3 units along the x-axis, followed by 4 units parallel to the y-axis, the position vector of the point $(3,4)$ may be represented by the vector $3\mathbf{i} + 4\mathbf{j}$. In general, $\mathbf{a} = a_1\mathbf{i} + a_2\mathbf{j}$, where a_1 and a_2 are components of **a** along the x- and y-axis respectively.

Any point (x,y) in the plane can be associated with the position vector $x\mathbf{i} + y\mathbf{j}$ and conversely $x\mathbf{i} + y\mathbf{j}$ is the position vector of the point (x,y).

Magnitude of a vector

Using this notation, the magnitude of the vector $x\mathbf{i} + y\mathbf{j}$ is $\sqrt{(x^2 + y^2)}$, by Pythagoras' theorem. Further, if $\mathbf{k}$ is the unit vector in the positive direction of the z-axis, then any point (x,y,z) in space has position vector $x\mathbf{i} + y\mathbf{j} + z\mathbf{k}$. The magnitude of this vector is $\sqrt{(x^2 + y^2 + z^2)}$.

Example 1. *If the position vectors of points P and Q are* $2\mathbf{i} + 3\mathbf{j}$ *and* $\mathbf{i} - 3\mathbf{j}$ *respectively, find the vector* $\overrightarrow{PQ}$.

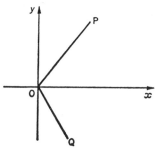

Fig. 44

Since
$$\overrightarrow{OQ} = \overrightarrow{OP} + \overrightarrow{PQ},$$
$$\overrightarrow{PQ} = \overrightarrow{OQ} - \overrightarrow{OP}$$
$$= (\mathbf{i} - 3\mathbf{j}) - (2\mathbf{i} + 3\mathbf{j})$$
$$= -\mathbf{i} - 6\mathbf{j}.$$

Alternatively, the vector which takes us from P to Q must change, by addition, $2\mathbf{i}$ into $+\mathbf{i}$ and $3\mathbf{j}$ into $-3\mathbf{j}$. Thus

$$\overrightarrow{PQ} = \mathbf{i} - 2\mathbf{i} + (-3\mathbf{j} - 3\mathbf{j})$$
$$= -\mathbf{i} - 6\mathbf{j}.$$

Example 2. *Find the position vector of M, the midpoint of OP, where OP = 2i + 3j.*

Since M is the midpoint of OP,

$$\overrightarrow{OM} = \tfrac{1}{2}\overrightarrow{OP} = i + \tfrac{3}{2}j.$$

Example 3. *If the position vectors of points A,B,C and D are* **a,b,c** *and* **d** *respectively, where* **a** = 2i + j, **b** = i − 3j, **c** = 4i *and* **d** = 3i − 4j, *show that AB is equal and parallel to CD.*

$$\overrightarrow{AB} = b - a$$
$$= -i - 4j$$

and

$$\overrightarrow{CD} = d - c$$
$$= -i - 4j.$$

Hence AB is equal and parallel to CD.

Example 4. **u** *and* **v** *are unit vectors inclined at 60°. Find the magnitude of 3***u** + 5**v**.

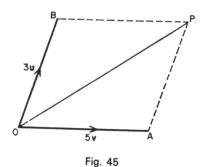

Fig. 45

If $\overrightarrow{OA} = 5v$ and $\overrightarrow{OB} = 3u$ and angle AOB = 60°, the vector sum is given by the diagonal OP of the parallelogram OAPB in Fig. 45.

Solution by drawing. Taking a scale of 2 cm to represent 1 unit, draw the triangle OAP, as in Fig. 46.

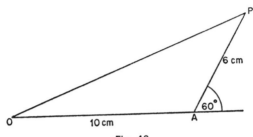

6 cm

60°

O   10 cm   A

Fig. 46

By measurement, OP = 14 cm. Hence the magnitude of the vector is 7.

Solution by calculation. Denoting OP by x and using the cosine formula,

$$x^2 = 3^2 + 5^2 - 2 \times 3 \times 5 \cos 120°$$
$$= 9 + 25 + 15 \quad \text{(since } \cos 120° = -0.5\text{)}$$
$$= 49$$
$$\therefore \quad x = 7.$$

The magnitude of $3\mathbf{u} + 5\mathbf{v}$ is therefore 7 units.

EXERCISE 10

1. Plot the following points and write down the position vectors of the lines joining the origin to each of them:

   (i) (3, 4)   (ii) (3, −5)   (iii) (4, −1),
   (iv) (−1, −2)   (v) (−1, 6)   (vi) (−1, 1).

2. Write down the coordinates of the points with the following position vectors:

   (i) $\mathbf{i} + \mathbf{j}$,   (ii) $-\mathbf{i} + 2\mathbf{j}$,   (iii) $-3\mathbf{i} - \mathbf{j}$,
   (iv) $\mathbf{i} - 5\mathbf{j}$,   (v) $-2\mathbf{i} - \mathbf{j}$,   (vi) $\mathbf{i} + 3\mathbf{j}$.

3. Draw two perpendicular axes and illustrate the following vectors:

   (i) $2\mathbf{i} + \mathbf{j}$,   (ii) $-\mathbf{i} - 2\mathbf{j}$,   (iii) $-3\mathbf{i} + \mathbf{j}$,   (iv) $\mathbf{i} - \mathbf{j}$.

4. If $\mathbf{a} = \mathbf{i} - 2\mathbf{j}$ and $\mathbf{b} = 3\mathbf{i} + \mathbf{j}$, find

   (i) $\mathbf{a} + \mathbf{b}$,   (ii) $\mathbf{a} + 2\mathbf{b}$,   (iii) $\mathbf{a} - \mathbf{b}$,   (iv) $\frac{1}{2}(\mathbf{a} + \mathbf{b})$.

5. If $\mathbf{u}$ and $\mathbf{v}$ are unit vectors inclined at 45°, find, by drawing, the magnitude of each of the following vectors:

   (i) $\mathbf{u} + \mathbf{v}$,   (ii) $\mathbf{u} - \mathbf{v}$,   (iii) $\mathbf{u} + 2\mathbf{v}$,   (iv) $3\mathbf{u} + \mathbf{v}$.

6. If $\mathbf{u}$ and $\mathbf{v}$ are unit vectors inclined at 30°, find, by drawing, the magnitude of each of the following vectors:

   (i) $\mathbf{u} + \mathbf{v}$,   (ii) $2\mathbf{u} - \mathbf{v}$,   (iii) $\mathbf{u} - 2\mathbf{v}$,   (iv) $2\mathbf{u} + 3\mathbf{v}$.

7. If $\mathbf{u}$ and $\mathbf{v}$ are unit vectors inclined at 40°, find, by drawing, the magnitude of (i) $2\mathbf{u} - \mathbf{v}$.
Deduce the magnitude of (ii) $\mathbf{u} - 2\mathbf{v}$,   (iii) $\mathbf{v} - 2\mathbf{u}$,   (iv) $4\mathbf{u} - 2\mathbf{v}$.

8. If $\mathbf{u}$ and $\mathbf{v}$ are unit vectors inclined at 120°, find, by calculation, the magnitude of $5\mathbf{u} + 8\mathbf{v}$.

9. If $\mathbf{u}$ and $\mathbf{v}$ are any unit vectors not in the same straight line, prove that $(\mathbf{u} + \mathbf{v})$ and $(\mathbf{u} - \mathbf{v})$ are perpendicular.

10. The position vectors of points A and B are $2\mathbf{i} + \mathbf{j}$ and $\mathbf{i} - 3\mathbf{j}$ respectively. Find the position vectors of
   (i) X, the midpoint of OA,
   (ii) Y, the midpoint of OB.
Verify that $\overrightarrow{XY} = \frac{1}{2}\overrightarrow{AB}$.

11. The position vectors of points A and B relative to the origin O are $3\mathbf{i} + \mathbf{j}$ and $-\mathbf{i} + 2\mathbf{j}$ respectively. Find the position vectors of
   (i) the point Y in OA such that OY : YA = 1 : 2,
   (ii) the point Z in OB such that OZ : ZB = 1 : 2.
Verify that $\overrightarrow{YZ} : \overrightarrow{AB} = 1 : 3$.

12. The position vectors of points A and B relative to the origin O are $3\mathbf{i} + 2\mathbf{j}$ and $\mathbf{i} + \mathbf{j}$ respectively. Find the position vectors of
   (i) the point P in OA produced such that OP : PA = 3 : −1.
   (ii) the point Q in OB produced such that OQ : QB = 3 : −1.
Find the ratio $\overrightarrow{AB} : \overrightarrow{PQ}$.

13. The position vectors $\mathbf{a}$ and $\mathbf{b}$ of points A and B relative to the origin are $\mathbf{i}$ and $\mathbf{j}$ respectively. Find a suitable value of k if

$$\tfrac{1}{6}\mathbf{i} + \tfrac{5}{6}\mathbf{j} = k\mathbf{a} + (1 - k)\mathbf{b}.$$

What can you deduce about the points A and B and the point X whose position vector is $\frac{1}{6}\mathbf{i} + \frac{5}{6}\mathbf{j}$?

14. The position vectors **a** and **b** of two points A and B relative to the origin are 2**i** and −4**j** respectively. Find k if

$$\tfrac{2}{3}\mathbf{i} - \tfrac{8}{3}\mathbf{j} = k\mathbf{a} + (1 - k)\mathbf{b}.$$

What can you deduce about A, B and the point X whose position vector is $\tfrac{2}{3}\mathbf{i} - \tfrac{8}{3}\mathbf{j}$?

15. A man starts from a point whose position vector relative to the origin O is $(\mathbf{i} + 2\mathbf{j})$ metres, and walks with velocity $(\mathbf{i} - \mathbf{j})$ metres per second. Find his position vector after (i) 1 s   (ii) 2 s   (iii) t s.

16. A man starts from a point P, position vector $(-300\mathbf{i} - 200\mathbf{j})$ metres and walks with velocity $(\mathbf{i} + 2\mathbf{j})$ metres per second, where **i** is a unit vector due east and **j** is a unit vector due north. Find
  (i) the speed at which he is walking,
 (ii) his position vector when he has walked for 50 s,
(iii) how long elapses until he is due north of O,
(iv) how long elapses until he is due west of O.

Section theorem and other geometrical results

The vector dividing the join of A and B in a given ratio

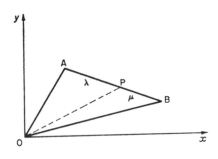

Fig. 47

Suppose that $\overrightarrow{OA}$ is represented by the vector **a** and $\overrightarrow{OB}$ by the vector **b**. What vector represents $\overrightarrow{OP}$, where P divides AB in the ratio $\lambda : \mu$?

$$\overrightarrow{OP} = \overrightarrow{OA} + \overrightarrow{AP}$$

$$= \mathbf{a} + \frac{\lambda}{\lambda + \mu}\ \overrightarrow{AB}$$

$$= \mathbf{a} + \frac{\lambda}{\lambda + \mu}\ (\mathbf{b} - \mathbf{a})$$

$$= \frac{\mu\mathbf{a} + \lambda\mathbf{b}}{\lambda + \mu}$$

So the vector representing $\overrightarrow{OP}$ is $\dfrac{(\mu\mathbf{a} + \lambda\mathbf{b})}{(\lambda + \mu)}$.

Deductions

1. The position vector of the mid-point of AB is $\frac{1}{2}(\mathbf{a} + \mathbf{b})$.
2. Putting $\mu = k$ and $\lambda = 1 - k$, any point of AB has position vector $k\mathbf{a} + (1 - k)\ \mathbf{b}$.
   $\mathbf{r} = k\mathbf{a} + (1 - k)\ \mathbf{b}$ is the vector equation of the straight line through points with position vectors $\mathbf{a}$ and $\mathbf{b}$.
3. The points with position vectors $\mathbf{a}$, $\mathbf{b}$ and $p\mathbf{a} + q\mathbf{b}$ are collinear if and only if $p + q = 1$.

The angle in a semicircle is a right angle

In question 9 of the previous exercise we saw that $\mathbf{u} + \mathbf{v}$ and $\mathbf{u} - \mathbf{v}$ are perpendicular if $\mathbf{u}$ and $\mathbf{v}$ are any two unit vectors not in the same straight line. This was most easily proved by saying that the diagonals of a rhombus are perpendicular.

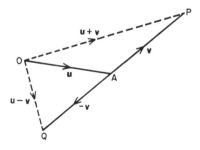

Fig. 48

Using this result in Fig. 48, since **u** and **v** are equal in magnitude a circle centre A will pass through O, P and Q. Thus angle POQ = 90°, for all positions of O, i.e. the angle in a semicircle is a right-angle.

There are many geometrical results which can be proved neatly using vectors.

The medians of a triangle are concurrent

Example 4. *If points A, B and C have position vectors* **a, b** *and* **c**, *find the position vector of G, the point on AD such that AG : GD = 2 : 1.*

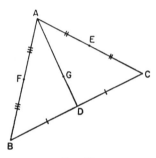

Fig. 49

Since D is the midpoint of BC, the position vector of D is $\frac{1}{2}(\mathbf{b} + \mathbf{c})$.

Since G divides AD in the ratio 2:1, by the Section theorem the position vector **g** of G is

$$\mathbf{g} = \frac{1}{2 + 1}\mathbf{a} + \frac{2}{2 + 1} \times \frac{1}{2}(\mathbf{b} + \mathbf{c})$$

$$= \tfrac{1}{3}(\mathbf{a} + \mathbf{b} + \mathbf{c}).$$

The symmetry of the results shows that G is also a point of trisection of the medians BE and CF. Thus the medians of any triangle are concurrent.

Example 5. *PQRS is a quadrilateral. Show that the lines joining the midpoints of the sides of PQRS form a parallelogram.*

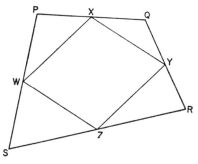

Fig. 50

If the position vectors of P, Q, R and S are **p**, **q**, **r** and **s**, then

the midpoint X of PQ has position vector $\frac{1}{2}(\mathbf{p} + \mathbf{q})$,
the midpoint Y of QR has position vector $\frac{1}{2}(\mathbf{q} + \mathbf{r})$,
the midpoint Z of RS has position vector $\frac{1}{2}(\mathbf{r} + \mathbf{s})$,
the midpoint W of SP has position vector $\frac{1}{2}(\mathbf{s} + \mathbf{p})$.

$$\overrightarrow{XY} = \tfrac{1}{2}(\mathbf{q} + \mathbf{r}) - \tfrac{1}{2}(\mathbf{p} + \mathbf{q})$$
$$= \tfrac{1}{2}(\mathbf{r} - \mathbf{p})$$
$$\overrightarrow{WZ} = \tfrac{1}{2}(\mathbf{r} + \mathbf{s}) - \tfrac{1}{2}(\mathbf{s} + \mathbf{p})$$
$$= \tfrac{1}{2}(\mathbf{r} - \mathbf{p})$$

Thus $\overrightarrow{XY}$ is equal and parallel to $\overrightarrow{WZ}$, i.e. XYZW is a parallelogram.

EXERCISE 11

1. Find the vector representing $\overrightarrow{AB}$, where the position vectors of A and B are $\mathbf{i} + 2\mathbf{j}$ and $3\mathbf{i} - \mathbf{j}$.
2. Find the position vector of the point of trisection nearer A of the line AB, where the position vectors of A and B are $\mathbf{i} + \mathbf{j}$ and $7\mathbf{i} + 4\mathbf{j}$ respectively.

3. Show that the points with position vectors $3\mathbf{i} + 4\mathbf{j}$, $\mathbf{i} - 2\mathbf{j}$, $\mathbf{i} + \mathbf{j}$ and $3\mathbf{i} + 7\mathbf{j}$ are the vertices of a parallelogram.

4. Show that the points with position vectors $\mathbf{i} - \mathbf{j}$, $2\mathbf{i}$, $\mathbf{i} + \mathbf{j}$ and $3\mathbf{i} + 3\mathbf{j}$ are the vertices of a trapezium in which one of the parallel sides is double the other.

5. The position vectors of A, B and C are $3\mathbf{i} + 2\mathbf{j}$, $2\mathbf{i} + 3\mathbf{j}$ and $3\mathbf{i} - 2\mathbf{j}$ respectively. D is the midpoint of BC and H has position vector $8\mathbf{i} + 3\mathbf{j}$. Show that $\overrightarrow{AH} = 2\overrightarrow{OD}$.

6. If A, B and C have position vectors $\mathbf{i} + \mathbf{j}$, $3\mathbf{i} + 2\mathbf{j}$ and $5\mathbf{i} - 4\mathbf{j}$ respectively, find the vector which represents the line joining A to the midpoint of BC.

In the following questions, points A, B, C and D have position vectors **a**, **b**, **c** *and* **d** *respectively.*

7. If E is the midpoint of AC, find the vector representing the line $\overrightarrow{BE}$.

8. What equation do **a**, **b**, **c**, and **d** satisfy if AC and BD bisect each other?

9. What is the vector condition that ABCD is a parallelogram?

10. Write down the vector representing EF, where E is the midpoint of AB and F is the midpoint of AC. What do you deduce?

11. If $\mathbf{c} - \mathbf{d} = \frac{1}{2}(\mathbf{b} - \mathbf{a})$, what can you say about the quadrilateral ABCD?

12. What is the condition that ABCD is a trapezium with AB parallel to DC?

13. If F is the midpoint of AB, what is the condition that AD is equal and parallel to FC?

14. In question 13, what is the condition that AD is parallel to FC?

15. BX is drawn equal and parallel to AC. Find the position vector of X.

16. BY is drawn parallel to AC and twice AC. Find the position vector of Y.

17. Show that the points with position vectors, **a**, $\mathbf{a} + \frac{2}{3}\mathbf{b}$, $\mathbf{a} + \mathbf{b}$ are collinear.

18. Show that only three of the points with position vectors **a**, $2\mathbf{a} + 3\mathbf{b}$, $3\mathbf{a} + 4\mathbf{b}$, $5\mathbf{a} + 8\mathbf{b}$ are collinear.

19. Show that the points with position vectors $\mathbf{a} + \mathbf{b}$, $3\mathbf{a} + 4\mathbf{b}$ and $\frac{3}{2}\mathbf{a} + \frac{5}{2}\mathbf{b}$ are collinear.

20. Find the value of λ if the points with position vectors $\lambda\mathbf{i}$, $2\mathbf{i} + 3\mathbf{j}$ and $3\mathbf{i} - 4\mathbf{j}$ are collinear.

21. Find the value of μ if the points with position vectors $\mu\mathbf{j}$, $\mathbf{i} - \mathbf{j}$ and $3\mathbf{i} + 4\mathbf{j}$ are collinear.

22. ABCD is a parallelogram. The line through the midpoint of AB parallel to BD meets AC at Y. Express the position vector of Y in terms of **a** and **c**.

23. ABCD is a parallelogram. The line joining the midpoint of AB to D meets AC in Z. Express the position vector of Z in terms of **a** and **c**.

24. In question 23, show that Z is a point of trisection of AC.

25. Show that a value of t can be found that makes $\frac{1}{2}t(\mathbf{a} + \mathbf{c}) + (1 - t)\mathbf{b}$ symmetrical in **a**, **b** and **c**. What can you deduce?

26. If the coordinates of P are (x, y) and the radius vector OP is rotated anti-clockwise through 90° about the origin O so that P moves to Q, what are the coordinates of Q?

27. Find the position vector of the centroid of the triangle whose vertices have position vectors $\mathbf{i} + \mathbf{j}$, $3\mathbf{i} + 2\mathbf{j}$ and $5\mathbf{i} - 4\mathbf{j}$.

28. D is the point of trisection of OA nearer O and E is the point of trisection of OB nearer B. BD and AE meet at P. Find the position vector of P.

[*Hint*: let OP $= \lambda\mathbf{a} + (1 - \lambda)\,\frac{2}{3}\mathbf{b}$ and also $\mu\frac{1}{3}\mathbf{a} + (1 - \mu)\mathbf{b}$.]

29. E is the midpoint of OB and D is the point of trisection of OA nearer O. BD and AE meet at P and OP meets AB at Q. Find the position vector of Q.

30. ABCD is a square of side $2a$. E and F are the midpoints of BC and CD, and EF meets AC in P. Take AB, AD as axes, and find the position vector of P in terms of **i** and **j**.

31. If **a**, **b**, **c**, **d**, are the position vectors of points A, B, C, and D in space show that $\frac{1}{4}(\mathbf{a} + \mathbf{b} + \mathbf{c} + \mathbf{d})$ lies on the line joining D to the centroid of the triangle ABC. What do you deduce?

32. (*Ceva's theorem*) If D, E, F are points on the sides of a triangle ABC such that BD : DC = 1 : λ, CE : EA = 1 : μ and AF : FB = 1 : ν, show that AD, BE, CF are concurrent if $\lambda\mu\nu = 1$.
[*Hint*: Write down the position vectors of D, E and F. Then find the position vector of the meet of AD and BE.)

33. (*Menelaus' theorem*) P, Q and R are points on the sides of the triangle ABC such that AP : PB = 1 : λ, BQ : QC = 1 : μ and CR : RA = 1 : ν. Show that P, Q, R are collinear if $\lambda\mu\nu = -1$.

CHAPTER 7

Applications of Vectors to Forces

Resultant of two forces

If two forces of P and Q act at an angle θ, draw OA to represent P and OB to represent Q. The resultant of the forces is OC where OACB is a parallelogram. This represents in magnitude and direction the single force which is equivalent to the two forces P and Q.

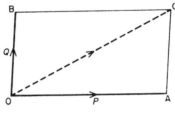

Fig. 51

Using the cosine formula,

$$OC^2 = OA^2 + AC^2 - 2OA.AC \cos OAC$$
$$= P^2 + Q^2 - 2PQ \cos(180° - \theta)$$
$$= P^2 + Q^2 + 2PQ \cos \theta.$$

The resultant of two forces P and Q which include an angle θ is therefore $(P^2 + Q^2 + 2PQ \cos \theta)^{\frac{1}{2}}$.

As a particular case of this, if the forces are at right angles, their resultant is $(P^2 + Q^2)^{\frac{1}{2}}$.

The parallelogram of forces

If two forces, acting at a point O, are represented in magnitude and direction by straight lines OA, OB, then their resultant is represented in magnitude and direction by the diagonal OC of the parallelogram OACB.

As an equation, this may be written

$$\overrightarrow{OC} = \overrightarrow{OA} + \overrightarrow{OB}.$$

It follows immediately that $\overrightarrow{CO}$ represents the single force which will keep the two forces $\overrightarrow{OA}$, $\overrightarrow{OB}$ in equilibrium. This is called the *equilibrant* of the two forces. So the forces $\overrightarrow{OA}$, $\overrightarrow{AC}$ and $\overrightarrow{CO}$ *acting at a point* will be in equilibrium.

The triangle of forces

If three forces acting on a particle are in equilibrium, then they may be represented graphically by the sides of a triangle taken in order. The converse is also true with the proviso that the forces must act at a point.

'Taken in order' means that the direction of the forces must follow each other round the triangle either in a clockwise or an anti-clockwise direction.

The triangle and the parallelogram of forces are equivalent theorems but the parallelogram does give the line of action of the forces which the triangle does not.

Three forces in equilibrium must meet at a point

It is fairly obvious that the resultant of two forces must pass through the point of intersection of those forces, from which it follows that three forces in equilibrium must meet in a point. A formal proof of this theorem is now given.

Suppose that three forces P, Q and R are in equilibrium. The forces P and Q are either parallel or meet in a point. If they are

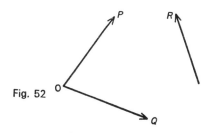

Fig. 52

parallel, we have already seen that the third force must also be parallel to them. If P and Q are not parallel, suppose they meet in a point O, as shown in Fig. 52. Since the forces are in equilibrium, they must have no moment about any point and in particular must have no moment about O. The moments of P and Q about O are each zero because they pass through O and so the moment of R about O must be zero. So either R must be zero or its line of action must pass through O. If R is zero, we have two forces only and so the line of action of R must pass through O.

The conclusion is that three forces in equilibrium must either be parallel or meet in a point.

N.B. It does not follow that four or more forces in equilibrium must meet in a point.

The triangle of forces provides a powerful method of dealing with three-force problems. For example, if we know that three forces P,Q,R acting at O are in equilibrium and we are given the magnitude of P and the directions of all three, we can find the magnitudes of Q and R by scale drawing.

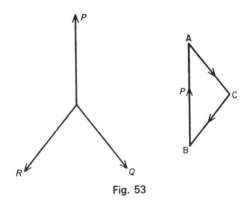

Fig. 53

Draw BA to represent P in magnitude and direction. Through A draw a line parallel to the direction of Q and through B draw another line parallel to the direction of R. If these two lines meet at C, then BC represents the force R and CA represents the force Q.

Lami's theorem

The triangle of forces can be expressed trigonometrically in the form of Lami's theorem which is often useful in calculation.

If three forces acting at a point are in equilibrium, then each is proportional to the sine of the angle between the other two.

In Fig. 53,

$$\frac{BC}{\sin A} = \frac{CA}{\sin B} = \frac{AB}{\sin C}.$$

But BC, CA, AB represent R, Q, P respectively and, since the sine of an angle is equal to the sine of its supplement, $\sin A = \sin \hat{POQ}$ etc. Therefore

$$\frac{Q}{\sin \hat{POR}} = \frac{R}{\sin \hat{POQ}} = \frac{P}{\sin \hat{ROQ}}.$$

Polygon of forces

The triangle of forces may be extended to cases in which any number of forces are acting.

If any number of forces acting on a body are in equilibrium, then they may be represented in magnitude and direction by the sides of a polygon taken in order.

If the forces act at a point and may be represented by the sides of a polygon taken in order, then they will be in equilibrium; if the lines of action of the forces are actually along the sides of the polygon they cannot be in equilibrium because all the forces will have moments which are either all clockwise or all anti-clockwise about any point inside the polygon as shown in Fig. 54.

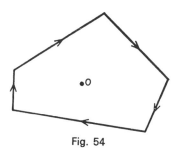

Fig. 54

Since the forces, if they acted at a point would be in equilibrium, they cannot tend to move the body in any one direction. Therefore the forces must be equivalent to a couple.

Deduction

Three forces acting along the sides of a triangle and proportional to these sides are equivalent to a couple. The moment of the couple may be found by taking moments about any one of the vertices or for that matter, about any other point.

Resultant of any number of forces

To find graphically the resultant of any number of forces, start at any point A and in any order draw lines AB, BC, CD, etc. to represent the given forces. If the last force is represented by FG, then the forces AB, BC, CD, DE, EF, FG and GA would be in equilibrium if they met at a point. Therefore GA represents the equilibrant of the forces and AG represents the resultant.

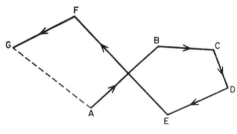

Fig. 55

N.B. The direction of the resultant is from the initial point to the final point.

AG represents the resultant in magnitude and direction but does not give its point of application. It is immaterial whether the polygon is re-entrant or not.

Example 1. *Two forces of 5 N and 6 N act at an angle of 120°. Find the magnitude of their resultant.*

$$R^2 = P^2 + Q^2 + 2PQ \cos \theta$$
$$= 5^2 + 6^2 + 2(5)(6) \cos 120°$$
$$= 25 + 36 + 2(5)(6)(-\tfrac{1}{2})$$
$$= 61 - 30$$
$$= 31$$

∴ $R = \sqrt{31}$ or 5.67 N.

Example 2. *A uniform bar AB of weight 20 N is hinged freely at A to a wall and is kept in a horizontal position by a string attached to the bar at B and inclined at an angle of 30° to BA. Find the magnitude and direction of the reaction at A.*

There are three forces acting on the bar; its weight, the tension in the string and the reaction at A. These forces must therefore meet in a point. Produce the line of action of the weight backwards

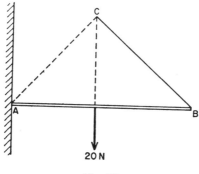

Fig. 56

as shown in Fig. 56 to meet the line of action of the tension at C. The force through A must be along AC and from the symmetry of the figure must make an angle of 30° with the horizontal.

The magnitude of the resultant may be found either by drawing or by calculation. By drawing, a separate triangle of forces is necessary. Draw a vertical line XZ on some scale to represent the weight, i.e. 20 N. Through Z draw a line at 30° to the horizontal to represent the reaction; through X draw a line at 30° to the

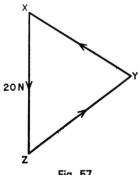

Fig. 57

horizontal to represent the tension in the string. If these lines meet at Y, as shown in Fig. 57, then XY represents the tension and YZ represents the reaction on the scale chosen for the weight.

By calculation apply Lami's theorem.

The angle between the weight and the tension is 60°.
The angle between the weight and the reaction is 60°.
The angle between the tension and the reaction is 120°.

$$\frac{20}{\sin 120°} = \frac{T}{\sin 60°} = \frac{R}{\sin 60°}.$$

But $\sin 120° = \sin 60°$. Hence

$$T = R = 20.$$

The reaction at A is 20 N and acts at 30° to the horizontal.

Example 3. *A rod AB which is 4 m long rests partly inside a smooth fixed hemispherical bowl of radius 2 m, at an angle of 30° to the horizontal. Find the distance of the centre of gravity of the rod from A.*

Fig. 58 shows the bowl drawn to a scale of 1 cm to 1 m. The common surface to the rod and bowl at A is along a tangent to the hemisphere and so the reaction is towards the centre. The common surface to the rod and the bowl at Y is along the rod

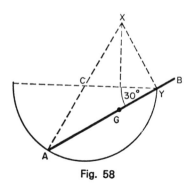

Fig. 58

and so the reaction to the rod there is perpendicular to the rod. Suppose that these two reactions meet at X. There are three forces acting on the rod; the two reactions and the weight of the rod. The weight of the rod must therefore act vertically down through X. Suppose this vertical meets the rod at G. Then G is the centre of gravity of the rod and AG is found by measurement to be 2.3 cm. The distance of the centre of gravity of the rod from A is therefore 2.3 m.

By calculation. Since AC = CY (radii), the angle YAC = 30°. Since the angle AYC = 30°, the angle CYX = 60°. Therefore the angle CXY also equals 60°. So CX = 2 m and AX = 4 m. Since CYX is an equilateral triangle, XY = 2 m.

Therefore AY = 2 tan 60° m and GY = 2 tan 30° m. Hence

$$AG = \left(2\sqrt{3} - \frac{2}{\sqrt{3}}\right) m$$

$$= \frac{4}{\sqrt{3}} m = 2.31 \text{ m}.$$

Example 4. *Forces of 10 N, 8 N, 6 N act along the sides BC, CA and AB respectively of an equilateral triangle ABC. Find their resultant in magnitude and direction.*

Take 1 cm to represent 2 N. Draw XY equal to 5 cm in length, YZ equal to 4 cm at an angle of 60° with XY, and ZT equal in length to 3 cm at an angle of 60° with YZ. Then XT represents

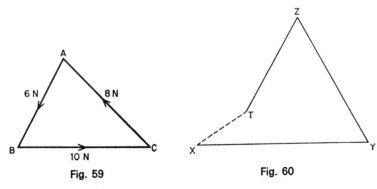

Fig. 59   Fig. 60

the resultant in magnitude and direction. By measurement
XT = 1.75 cm and so the resultant is 3.5 N. The angle the resultant makes with XY is 30°. A method of calculation will be given
in the next chapter.

EXERCISE 12

1. Forces of 3 N and 4 N meet at an angle of 60°. Find their resultant.
2. A particle is acted on by a force of 6 N due north, a force of 12 N
   due east and a force of 4 N due west. Find the resultant of these
   forces.
3. Two forces of 5 N and 12 N act at right angles to each other. Find
   their resultant.
4. Find the angle between two forces of 6 N and 14 N if the magnitude
   of their resultant is 14 N.
5. Two forces of 6 N and 9 N are inclined at an angle of 35°. What
   is their resultant?
6. Two forces act at right angles to each other. One of the forces is
   15 N and their resultant is 17 N. What is the magnitude of the
   other force?
7. Forces of 10 N due east and 8 N due north act on a body. What
   is the direction of their resultant?
8. If the resultant of two forces of 10 N and 12 N is 17 N, find the
   angle between the forces.
9. Two forces of 10 N and 16 N make an angle of 50° with each other.
   Find the magnitude of their resultant.
10. Two horizontal cables in which the tensions are 900 N each, pull a
    boat along a canal. If the cable makes angles of 20° with the canal
    banks, find the resultant pull on the boat.

11. Forces of 5 N, 6 N, 7 N and 8N act along the sides AB, BC, CD and DA respectively of a rectangle. Find the magnitude of their resultant.

12. Three forces each of 10 N act along the sides AB, BC and CA of an equilateral triangle. What is their resultant?

13. Forces of 10 N, 12 N and 14 N act along the sides AB, BC and CA of an equilateral triangle. What is the magnitude of their resultant?

14. Find the resultant in magnitude and direction of forces of 10 N due east, 8 N due north and 12 N acting in a direction 030°.

15. Forces of magnitudes 30 N, 40 N and 50 N act on a particle and are in equilibrium. Show that the lines of action of two of them are perpendicular.

16. Show that the resultant of two equal forces inclined at 120° is another equal force.

17. Find the angle between two forces F and $2F$ if their resultant is $F\sqrt{3}$.

18. ABCD is a square. Forces of 10 N, 20 N, 30 N, 30 N, 40 N act along AB, BC, CD, DA, AC respectively. Find their resultant in magnitude and direction.

19. ABCDEF is a regular hexagon. Forces of 10 N, 30 N, 50 N, 60 N act along AB, BC, DE, EF respectively. Find their resultant in magnitude and direction.

20. A uniform bar AB of weight 4 N and length 10 m is hinged at A and is kept in a horizontal position by a string attached at B and inclined at an angle of 30° with BA. By taking moments about A, find the tension in the string. Using the triangle of forces, find the magnitude and direction of the reaction at A.

21. A particle of weight 6 N is kept at rest on a smooth plane inclined at 30° to the horizontal by a force which makes an angle of 20° with the upward line of greatest slope. Find the magnitude of this force.

22. A particle of weight 4 N is attached to the lower end of a string 5 m in length the upper end of which is fixed. It is pulled a distance 2 m from the vertical through the fixed end by a horizontal force. Find the magnitude of the horizontal force.

23. A particle of weight 20 N is suspended by two strings of lengths 2.5 m and 2 m the upper ends of which are fixed at two points on the same horizontal level and 4 m apart. Find the tensions in the strings.

24. A ring of weight 4 N can slide freely on a string of length 4 m, whose ends are fixed at two points on the same horizontal level and 3 m apart. Find the tension in the string in the position of equilibrium.

25. A weight is supported by two strings inclined at 50° to each other. If the tensions in the strings are 4 N and 8 N, find the weight of the body.

26. A sphere of radius 8 cm rests against a smooth vertical wall and is supported by a string of length 16 cm which is attached to the wall and to the surface of the sphere. If the sphere weighs 6 N, find the tension in the string and the angle it makes with the horizontal.

27. A uniform rod AB of weight 8 N and length 2 m is supported at an angle of 30° with the horizontal and with B below A by a hinge at A and a string BT. If the angle ABT is 50°, find the tension in the string and the reaction at the hinge.

28. A uniform bar AB of length 3.6 m and weight 10 N is supported by two strings AC and BC of lengths 3 m and 4.5 m, where C is fixed. Find the inclination of the bar to the horizontal and the tensions in the strings in the position of equilibrium.

29. A ladder of weight 45 N and length 6 m rests in equilibrium against a smooth vertical wall and with its foot on rough horizontal ground. The foot of the ladder is 3.6 m from the wall and the centre of gravity of the ladder is 2.4 m from its lower end. Find the reaction at the wall and the magnitude and direction of the reaction at the ground.

30. A bar AB is supported in a horizontal position by two strings AT and BT. Given that the angle BAT = 60° and that the angle ABT = 45°, find the position of the centre of gravity of the bar.

31. A bar AB, of length 3.6 m and weight 8 N has its centre of gravity 1.5 m from A. The bar hangs horizontally supported by two strings AT and BT. Given that angle BAT = 30°, find the angle ABT and the tension in the two strings.

32. A uniform bar AB of weight W, hinged at A, is kept at an angle 2α above the horizontal by a string BC, where C is a point vertically above A and such that AC = AB. Find the tension in the string and the magnitude and direction of the reaction at A.

33. A mass of 5 kg is supported by two strings making angles of 20° and 30° with the vertical. Find, graphically or otherwise, the tensions in the strings.

If the strings are adjusted so that one of them is at 50° to the vertical and has a tension of $3g$ N, find the tension and direction of the other string.   (O. & C.*)

34. A non-uniform rod is suspended by two strings AB, DC from two points A, D at the same horizontal level. AB = 15 cm, BC = 20 cm, CD = 10 cm, AD = 30 cm and angle BAD = 60°. The centre of gravity of the rod is at G. Find the length BG.

If the weight of the rod is 100 N, find the tensions in AB and CD.   (O. & C.*)

REVISION EXERCISE ON VECTORS

1. A man is initially 3 km due east of a point O, and walks at 5 km h^{-1} north-east. How far is he from O 2 hours after he started?

2. A particle has initial velocity of 3 m s^{-1} due east, and an acceleration of 5 m s^{-2} north-east. What is the magnitude of the velocity of the particle after 2 s?

3. A body mass 4 kg has initially a velocity of 3 m s^{-1} due east, and is acted on by a force of 20 N, in a direction north-east. What is the magnitude of the velocity of the body after the force has been acting for 2 s?

4. A particle has a velocity of 3 m s^{-1} due south, and has an acceleration of 1 m s^{-2} south-west. What is the magnitude of the velocity after 10 s?

5. A particle has an initial velocity of 3 m s^{-1} due west, and has an acceleration of 1 m s^{-2} south-east. Find the magnitude of the velocity of the particle after 10 s.

6. At 9.00 a.m. a man X is 4 km north 20° east of an observer O; he walks south-east at 5 km h^{-1}. Find when X is due east of O.

7. A body mass 20 kg initial velocity 3 m s^{-1} due east is acted on by a force of 4 N, in the direction N 20° W. Find the magnitude and direction of the velocity of the body after 30 s.

8. Three points A, B and C are such that AB = 4 cm, BC = 5 cm and angle ABC = 50°. A particle travels from A to C along AC in 5 s. Find the average speed of the particle.
   If the particle starts from rest and travels with variable acceleration, what is the least acceleration it must have at some one instant to complete the journey in 5 s?

9. If the particle in question 8 has, instead, an initial velocity of 0.5 cm s^{-1}, what is the least acceleration it now must have at some one instant, in order to complete its journey in 5 s?

10. At a certain instant a body has a velocity of 3 m s^{-1} eastward and a constant acceleration of 2.5 m s^{-2} in a direction N 30° E. Find, graphically or otherwise, the magnitude and direction of its velocity two seconds later. When the body is moving in a direction due north-east, find the magnitude of its velocity. (M.E.I.)

11. A particle has initial velocity $3\mathbf{i}$ m s^{-1}, and an acceleration $(\mathbf{i} + 2\mathbf{j})$ m s^{-2}. Find the magnitude of the velocity (a) after 5 s, (b) after t seconds.

12. A particle has initial velocity of $(\mathbf{i} + \mathbf{j})$ m s^{-1}, and an acceleration of $(\mathbf{i} + 2\mathbf{j})$ m s^{-2}. Find
    (a) the magnitude of the velocity after t seconds,
    (b) after how many seconds the speed of the particle is 17 m s^{-1}.

13. A particle has initial velocity of $-\mathbf{i}$ m s^{-1} and an acceleration of $(\mathbf{i} + 4\mathbf{j})$m s^{-2}. Find
    (a) after how many seconds the speed of the particle is 41 m s^{-1},
    (b) after how many seconds the velocity of the particle is parallel to $\mathbf{j}$.

14. At 11.00 a.m. the position vector of an aircraft relative to an airport O is $(200\mathbf{i} + 30\mathbf{j})$ km, $\mathbf{i}$ and $\mathbf{j}$ being unit vectors east and north respectively. The velocity of the aircraft is $(180\mathbf{i} - 120\mathbf{j})$ km h^{-1}. Find
    (a) when the aircraft is due east of O,
    (b) how far it then is from O,
    (c) how far it is from O at 12.00 noon.

15. A travelling crane hoists a load from relative rest with a vertical acceleration of 0.03 m s^{-2} while travelling horizontally with a uniform velocity of 0.9 m s^{-1}. Find the velocity $\mathbf{v}$ of the load at time t in terms of a unit vector $\mathbf{i}$ in the direction of horizontal travel and a unit vertical vector $\mathbf{j}$. What is the speed of the load after 40 seconds?   (S.M.P.)

16. Two forces magnitudes 8 N and 10 N act at a point. The angle between their lines of action is 50°. By drawing a suitable parallelogram, find the magnitude of the single force to which they are equivalent.

17. Two strings are attached to a nail. The tension in one string is 7 N and in the other string is 12 N, and the angle between the strings is 60°. Find the magnitude of the single force to which the two tensions are equivalent.

18. Two boys are paddling a canoe. One exerts a force of 100 N at 40° to the direction of motion of the canoe: the paddle of the second boy is in the water on the other side of the canoe and he exerts a force of 120 N at 30° to the direction of motion. To what single force are these equivalent? What is the angle between this force and the direction of motion of the canoe?

19. Three dogs are fighting over a large bone. If one dog exerts a force of $(100\mathbf{i} + 150\mathbf{j})$ N, and the second dog exerts a force of $(200\mathbf{i} + 50\mathbf{j})$ N, find the vector which describes the force exerted by the third dog, if the bone does not move.

20. If $\mathbf{r}_1$ is the position vector of a particle and $\mathbf{r}_2$ its position vector after time t, define the velocity $\mathbf{v}$ in terms of $\mathbf{r}_1$, $\mathbf{r}_2$ and t.

21. If $\mathbf{a} = 3\mathbf{i} + 5\mathbf{j}$, $\mathbf{b} = \mathbf{i} - 2\mathbf{j}$ and $\mathbf{c} = 5\mathbf{i} + \mathbf{j}$, find scalars λ and μ such that $\mathbf{c} = \lambda\mathbf{a} + \mu\mathbf{b}$.

22. If $\mathbf{a} = \mathbf{i} - 2\mathbf{j}$, $\mathbf{b} = 3\mathbf{i} + \mathbf{j}$ and $\mathbf{c} = -\mathbf{i} - 5\mathbf{j}$, find scalars λ and μ such that $\mathbf{c} = \lambda\mathbf{a} + \mu\mathbf{b}$.

23. If $\mathbf{a} = \mathbf{i} - \mathbf{j}$, $\mathbf{b} = -2\mathbf{i} + 2\mathbf{j}$ and $\mathbf{c} = 5\mathbf{i} + 4\mathbf{j}$, why is it impossible to find scalars λ and μ such that $\mathbf{c} = \lambda\mathbf{a} + \mu\mathbf{b}$?

24. Prove that, in general, if P is any point in the plane OAB, scalars λ and μ can be found such that $\overrightarrow{OP} = \lambda\overrightarrow{OA} + \mu\overrightarrow{OB}$. What is the exceptional case?

25. i and j are unit vectors in the Ox and Oy directions respectively. Show in a diagram the displacement vector $3\mathbf{i} + 4\mathbf{j}$, and find its magnitude and direction. Find a displacement vector **b** of equal magnitude but in a direction at $90°$ anti-clockwise to that of **a**. Express the displacement vector $\mathbf{c} = 24\mathbf{i} + 7\mathbf{j}$ as $p\mathbf{a} + q\mathbf{b}$ where p and q are numbers to be found. Hence calculate the magnitude of the displacement vector **c** in two different ways. (M.E.I.)

26. Interpret geometrically

$$\mathbf{a} - \mathbf{b} = \mathbf{c} - \mathbf{d} \Leftrightarrow \tfrac{1}{2}(\mathbf{a} + \mathbf{d}) = \tfrac{1}{2}(\mathbf{b} + \mathbf{c})$$

where **a**, **b**, **c** and **d** are position vectors of four coplanar points A, B, C and D respectively.

27. Show that the vector equation $\mathbf{p} + \mathbf{r} = \mathbf{q} + \mathbf{s}$, where **p** is the position vector of a point P, and so on, is equivalent to each of the statements

PQ is equal and parallel to SR;
PS is equal and parallel to QR.

PQRS, P*Q*R*S* are parallelograms; A is the midpoint of PP*, B is the midpoint of QQ*, C of RR* and D of SS*. Prove that ABCD is also a parallelogram.
Will the same result be true if PQRS, P*Q*R*S* are not coplanar? Give reasons. (S.M.P.)

28. The position vectors of points A, B, C and D are **a**, **b**, **c** and **d** respectively. The midpoint of AB is X, of BC is Y, of CD is Z and of DA is W. Prove by vector methods that XYZW is a parallelogram.

29. In the triangle AOB, $AO = OB$. P is any point on the internal bisector of angle AOB. Denoting $\overrightarrow{OA}$ by **a**, $\overrightarrow{OB}$ by **b** and $\overrightarrow{OP}$ by **r**, express **r** in terms of **a**, **b** and a parameter λ.

30. P is any point on the internal bisector of any angle XOY. Denoting $\overrightarrow{OX}$ by **x**, $\overrightarrow{OY}$ by **y** and $\overrightarrow{OP}$ by **r**, express **r** in terms of **x**, **y**, x and y, where x and y are the magnitudes of the vectors **x** and **y** respectively.

CHAPTER 8

Components and Resolution: Equivalent Systems of Forces

Components

If, in Fig. 61(a), $\overrightarrow{AB}$ represents a force in magnitude and direction and ACBD is any parallelogram, then by vectors,

$$\overrightarrow{AB} = \overrightarrow{AC} + \overrightarrow{CB}$$
$$\Rightarrow \overrightarrow{AB} = \overrightarrow{AC} + \overrightarrow{AD}.$$

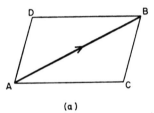
(a)

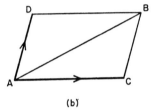
(b)

Fig. 61

So the force $\overrightarrow{AB}$ may be replaced by the two forces $\overrightarrow{AC}$ and $\overrightarrow{AD}$. Now suppose that we make AC horizontal and AD vertical and get the rectangle shown in Fig. 62(a). Then $\overrightarrow{AB}$ may be replaced by $\overrightarrow{AC}$ horizontally and $\overrightarrow{AD}$ vertically. Since a vertical force has

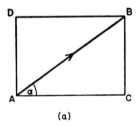
(a)

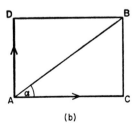
(b)

Fig. 62

not any tendency to move a body in a horizontal direction, $\overrightarrow{AC}$ is the effective horizontal contribution of the force $\overrightarrow{AB}$ and similarly $\overrightarrow{AD}$ is the effective vertical contribution of the force $\overrightarrow{AB}$. These are called the horizontal and vertical *components* of the force $\overrightarrow{AB}$. If the angle CAB = α and the force $\overrightarrow{AB}$ is F, then since AC = AB cos α, the horizontal component of F is $F \cos \alpha$ and similarly the vertical component of F is $F \sin \alpha$.

To find the component of a force in any given direction, multiply the force by the cosine of the angle between the force and the direction into which the force is being resolved.

By this rule, the vertical component of F in Fig. 62 is $F \cos (90° - \alpha)$ or $F \sin \alpha$ as already seen. We are now assuming that the force is being resolved into two perpendicular components.

If it is required to resolve a force F into a direction which makes an angle $(180° - \theta)$ with it, then the component is $F \cos (180° - \theta)$ or $-F \cos \theta$ i.e. the component in that direction is negative.

Example 1. *Suppose a force of 6 N acts at an angle of 60° with Ox as shown in Fig. 63. Find the components of the force in the directions (i) Ox; (ii) Ox'; (iii) Oy; (iv) Oy'.*

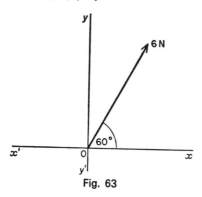

Fig. 63

(i)   The component along Ox is 6 cos 60° N.
(ii) The component along Ox' is 6 cos 120° N = −6 cos 60° N.
(iii) The component along Oy is 6 cos 30° N = 6 sin 60° N.
(iv) The component along Oy' is 6 cos 150° N = −6 cos 30° N
      = −6 sin 60° N.

Example 2. *Suppose a toy train (Fig. 64) is pulled along a track by a horizontal force of 10 N, making an angle of 25° with the track. Find the component of this force along the track.*

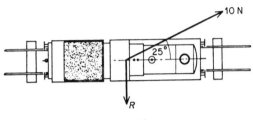

Fig. 64

The resolved part of the force along the track is 10 cos 25° N and this is the effective force pulling the train along the track. The other component perpendicular to the track is 10 sin 25° N and this is balanced by the lateral reaction of the track, i.e. $R = 10 \sin 25°$ N.

Example 3. *Suppose a mass m lies on a smooth plane inclined at an angle α to the horizontal. The forces acting on the mass are its weight mg and the normal reaction R. Find the acceleration of the mass down the plane.*

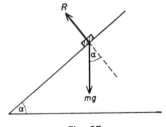

Fig. 65

The resolved part of the weight perpendicular to the slope is $mg \cos α$ and since the mass does not move in a direction perpendicular to the slope, $R = mg \cos α$.

The other component of the weight down the slope is $mg \sin \alpha$ and this force accelerates the body down the slope.

$$F = ma$$
$$\Rightarrow mg \sin \alpha = ma$$
$$\Rightarrow a = g \sin \alpha.$$

Forces in Equilibrium

If any number of forces acting on a body are in equilibrium then there is no tendency for the body to move in any given direction, or to turn about any point. Therefore the algebraic sum of all the forces in any given direction must be zero as well as the sum of the moments of all the forces being zero. It will be found in practice that the directions most useful for resolution are either horizontal and vertical, or along a given plane and perpendicular to that plane.

Similarly to find the resultant of any number of forces, the components of the resultant in any two perpendicular directions must be equal to the algebraic sums of the components of all the forces in these two directions. If the sums of the components in these perpendicular directions are X and Y, then the resultant is $\sqrt{(X^2 + Y^2)}$ and, from Fig. 66, the resultant will make an angle α, where $\tan \alpha = Y/X$, with the direction of X. This gives an alternative to the polygon of forces for finding the resultant of any number of forces.

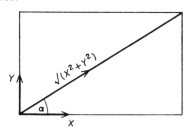

Fig. 66

Resultant of two forces

If two forces P and Q include an angle α, we have already seen that their resultant is $\sqrt{(P^2 + Q^2 + 2PQ \cos \alpha)}$.

This may be proved also by the method of resolution which, in addition, gives easily the direction of the resultant.

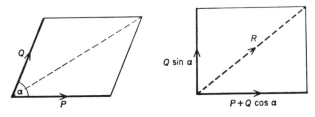

Fig. 67

The force Q may be replaced by $Q \cos \alpha$ in the direction of P and $Q \sin \alpha$ perpendicular to it. We then have two perpendicular forces $(P + Q \cos \alpha)$ and $Q \sin \alpha$. The resultant of these two perpendicular forces is given by

$$R^2 = (P + Q \cos \alpha)^2 + (Q \sin \alpha)^2$$

i.e. $$R^2 = P^2 + Q^2 + 2PQ \cos \alpha$$

since $\cos^2 \alpha + \sin^2 \alpha = 1$.

The angle the resultant makes with the direction of P is

$$\tan^{-1} \frac{Q \sin \alpha}{P + Q \cos \alpha}.$$

N.B. If the forces P and Q are perpendicular, the angle the resultant makes with the direction of P is $\tan^{-1} (Q/P)$, and the magnitude of the resultant is $\sqrt{(P^2 + Q^2)}$.

The resolved part of P in its own direction is P. The resolved part of P perpendicular to itself is $P \cos 90°$ or 0.

Example 4. (*solved graphically in Chapter 6*).
Forces of 10 N, 8 N and 6 N act along the sides BC, CA, and AB respectively of an equilateral triangle ABC. Find their resultant in magnitude and direction.

The resolved part of 10 N along BC is 10 N.
The resolved part of 8 N along BC is $-8 \cos 60°$ N or -4 N.
The resolved part of 6 N along BC is $-6 \cos 60°$ N or -3 N.

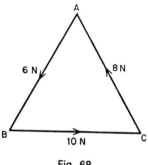

Fig. 68

The resolved part of the resultant along BC is $(10 - 4 - 3)$ N or 3 N.

The resolved part of 10 N perpendicular to BC is 0.

The resolved part of 8 N perpendicular to BC upwards is 8 sin 60° N or $4\sqrt{3}$ N.

The resolved part of 6 N perpendicular to BC upwards is -6 sin 60° N or $-3\sqrt{3}$ N.

The resolved part of the resultant perpendicular to BC upwards is $(4\sqrt{3} - 3\sqrt{3})$ N or $\sqrt{3}$ N.

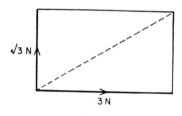

Fig. 69

The resultant is $\sqrt{(3 + 9)}$ N or $\sqrt{12}$ N, i.e. 3.46 N.
The angle the resultant makes with BC is $\tan^{-1}(\sqrt{3}/3)$ or $\tan^{-1}(1/\sqrt{3})$ or 30°.

EXERCISE 13

1. Find the components of the given force in the directions Ox and Oy in each diagram in Fig. 70.

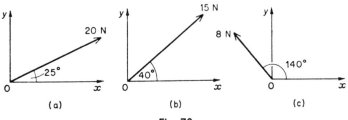

Fig. 70

2. In each of the following find the magnitude of the resultant of the forces given, and the angle the resultant makes with Ox, where Ox and Oy are perpendicular axes.

(a) 4 N along Ox, 5 N along Oy,
(b) 2 N along Ox, 7 N along Oy,
(c) -6 N along Ox, 3 N along Oy,
(d) -6 N along Ox, -4 N along Oy,
(e) 1 N along Ox, -5 N along Oy.

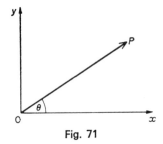

Fig. 71

3. In each of the following the data refers to Fig. 71. In addition, X and Y are forces along Ox and Oy respectively. Find the components of the force P along Ox and Oy, and hence the magnitude of the resultant of each system of forces, and the angle it makes with Ox.

(a) $X = 2$ N, $Y = 0$, $P = 3$ N, $\theta = 30°$.
(b) $X = 0$, $Y = 3$ N, $P = 4$ N, $\theta = 30°$.

(c) $X = 2$ N, $Y = 3$ N, $P = 4$ N, $\theta = 40°$.
(d) $X = -2$ N, $Y = 3$ N, $P = 4$ N, $\theta = 50°$.
(e) $X = -2$ N, $Y = -1$ N, $P = 6$ N, $\theta = 64°$.
(f) $X = 2$ N, $Y = 3$ N, $P = 5$ N, $\theta = 120°$.
(g) $X = -4$ N, $Y = 2$ N, $P = 1$ N, $\theta = 200°$.
(h) $X = 5$ N, $Y = 1$ N, $P = 3$ N, $\theta = 310°$.

4. In each of the following parts of this question the data refers to Fig. 72. In addition, X and Y are forces along Ox and Oy respectively. The magnitude of any force not listed in any one part is 0 in that part. Find the magnitude of the resultant of the systems of forces given in each part.

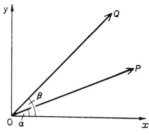

Fig. 72

(a) $X = 2$ N, $P = 3$ N, $Q = 4$ N, $\alpha = 20°$, $\beta = 40°$.
(b) $X = -2$ N, $P = 3$ N, $Q = 5$ N, $\alpha = 30°$, $\beta = 50°$.
(c) $X = 2$ N, $Y = 4$ N, $P = 3$ N, $Q = 4$ N, $\alpha = 10°$, $\beta = 40°$.
(d) $X = 1$ N, $Y = -3$ N, $P = 3$ N, $Q = 4$ N, $\alpha = 80°$, $\beta = 115°$.

Equivalent systems of forces

Looking back at Fig. 61, the forces $\overrightarrow{AC}$ and $\overrightarrow{AD}$ could be replaced by a single force $\overrightarrow{AB}$, and these two systems of forces are said to be equivalent. Similarly in Fig. 73(a) the forces 6 N and 8 N are equivalent to a force 10 N in Fig. 73(b). The resolved parts of the forces in any direction in the two systems are equal, and the two systems have the same moments about any point. Notice though that if the 6 N along AD is replaced by 6 N along BC, as in Fig. 73(c), these systems are not equivalent. The components in any direction are still equal, but the moments about any point are not equal. In particular, taking moments about O,
    the moment of system (a) is 0,
    the moment of system (c) is 24 N m.

Systems (a) and (c) are equivalent to forces equal in magnitude and direction, but acting along different lines.

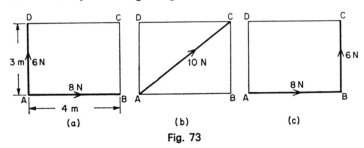

Fig. 73

It cannot be too strongly emphasized that in investigating whether two systems of forces are equivalent, or in finding a second system of forces equivalent to an existing system, it is most important to draw two diagrams, one for each system of forces.

Example 5. *Forces 1 N, 3 N and 3 N act along the sides of a square ABCD, side 1 m as in Fig. 74(a). Find the single force to which they are equivalent and the position of the point X in which the line of action of that force meets AB.*

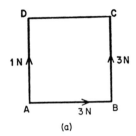 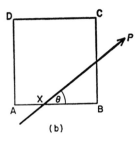

Fig. 74

The resolved parts of the two systems along AB being equal, in Fig. 74(b)

$$3 = P \cos \theta.$$

The resolved parts of the systems parallel to AD being equal,

$$4 = P \sin \theta,$$

Thus

$$3^2 + 4^2 = P^2 \cos^2\theta + P^2 \sin^2\theta,$$

hence

$$P = 5$$

and

$$\tan \theta = \tfrac{4}{3}$$

i.e.

$$\theta \simeq 53° \, 8'$$

Denoting the length of AX by x, and equating the moments about X of the two systems,

$$1.x - 3(1 - x) = 0,$$
$$x = 0.75,$$

i.e., X is 0.75 m from A.

The forces are therefore equivalent to a force of 5 N at an angle with AB whose tangent is $\tfrac{4}{3}$, through a point in AB 0.75 m from A.

EXERCISE 14

Questions 1 — 3 refer to a rectangular framework ABCD in which AB = 4 m, BC = 3 m.

1. Find the magnitude and direction of the single force equivalent to

6 N acting from A to B,
2 N acting from B to C,
4 N acting from C to D,
8 N acting from A to D.

Find also the distance from A of the point in which the line of action of the force meets AB.

2. Find the magnitude and direction of the single force equivalent to

6 N acting from A to B,
4 N acting from B to C,
10 N acting from C to A.

Find the line of action of this force.

3. Find the magnitude and direction of the single force equivalent to

6 N from A to B,
6 N from B to C,
15 N from C to A,
5 N from D to B.

Find also the distance from A of the point at which the line of action of this force meets BA produced.

4. Forces 3 N, 5 N, 2 N and 1 N act along the sides AB, BC, CD and DA of a square. If these forces are equivalent to forces X along AB, Y along BC and Z along CD, find the values of X, Y and Z.

5. Forces 1 N, 2 N and 1 N act along AB, BC, CD, three of the sides of the square ABCD side a. Find the magnitude and direction of the single force to which these are equivalent and the distance from A of the point at which its line of action cuts AB.

6. The coordinates of the vertices of the square ABCD are (0,0), (2,0), (2,2) and (0,2) respectively. Forces of magnitude 2 N act along AB, CD and AD. Find the single force to which these are equivalent, and the equation of its line of action.

7. In the triangle ABC, AB = AC = 8 cm and BC = 3 cm. Forces of 16 N act along AB and CA and a force of 6 N acts along BC. Show that these forces are equivalent to a couple, and are not in equilibrium. Find the magnitude of the couple.

8. Forces 2 N, 3 N and 4 N act along the sides AB, BC and CA of an isosceles triangle ABC in which angle ABC = 90°. Find the magnitude of the force to which these are equivalent, and the angle which the line of action of that force makes with AB.

9. Forces $(3i + 4j)$ N, $(i - 3j)$ N and $(2i - j)$ N act on a particle. Find the single force to which these are equivalent.

10. Three points on a lamina A, B and C have position-vectors $4i$, $4i + 3j$ and $3j$ respectively. A force of $6j$ N is applied at A, a force of $4i$ N is applied at B and a force $(8i - 6j)$ N at C. Find the single force to which these are equivalent and the position-vector of the point at which its line of action meets AB.

11. Find the resultant in magnitude and direction of forces 4 N due east, 2 N north-east, and 3 N in a direction N 30°W.

12. ABCD is a square. Forces of 2 N, 3 N, 4 N, 2 N and 5 N act along AB, BC, CD, DA and AC respectively. Find their resultant in magnitude and direction.

13. Forces of 2 N, 3 N, 4 N and 5 N act along the sides AB, BC, CD DA of a square. Find their resultant in magnitude and direction.

14. Forces of 2 N, 3 N, 5 N act along the sides AB, BC and CA of an equilateral triangle. Find their resultant in magnitude and direction.

15. Forces F, $3F$, $2F$ and $5F$ act along the sides AB, BC, CD and DA of a square ABCD. Find their resultant in magnitude and direction.

16. D is the midpoint of the side BC of a triangle ABC. Find the resultant in magnitude and direction of forces represented by AB, BC, CA and AD.

17. If G is the intersection of the medians of the triangle ABC, prove that forces represented by $\overrightarrow{GA}$, $\overrightarrow{GB}$ and $\overrightarrow{GC}$ are in equilibrium.
(*Hint*: $\overrightarrow{AB} = \overrightarrow{AG} + \overrightarrow{GB}$)

18. ABCDEF is a regular hexagon. Forces of 1 N, 2 N, 5 N, 6 N act along AB, CD, FE and AF respectively. Find their resultant in magnitude and direction.

19. G is the intersection of the medians of a triangle ABC. Prove that the resultant of forces represented by $6\overrightarrow{GA}$, $6\overrightarrow{GB}$ and $3\overrightarrow{GC}$ is a force represented by $2\overrightarrow{CD}$, where D is the mid-point of AB.

20. Forces of 4 N, 6 N and 8 N act along the sides AB, BC and CA of an equilateral triangle. Show that the resultant is perpendicular to BC and find its magnitude.

21. A uniform telephone pole weighing 4000 N is being raised by men on the ground pulling at a rope tied to the top with the foot of the pole resting without slipping on the ground. Find the tension in the rope when the pole is inclined at 60° to the horizontal and the rope makes an angle of 30° with the ground.

22. Forces represented by $\overrightarrow{AB}$, $\overrightarrow{BC}$, $2\overrightarrow{CA}$ act along the sides of an equilateral triangle ABC. Find the magnitude and direction of their resultant.

23. A uniform rod rests against a smooth vertical wall with its other end on a smooth plane inclined at an angle x with the horizontal. If the rod makes an angle y with the horizontal, show that $\tan x . \tan y = \frac{1}{2}$.

24. A uniform ladder stands on a smooth floor and leans against a smooth wall, being held at an angle x with the vertical by a horizontal rope which joins the foot of the ladder and the nearest point of the wall. If the weight of the ladder is W, show that the tension in the rope is $\frac{1}{2}W \tan x$.

25. ABC is a triangle. It is given that AB = 30 cm, BC = 40 cm and CA = 50 cm. Forces of 6 N, 8 N and 10 N act among AB, BC, CA respectively. Show that the forces are equivalent to a couple and find its moment.

26. A uniform bar BC is suspended from a peg A by strings AB and AC. The angle ABC is 30° and ACB is a right angle. Find the angle the bar makes with the horizontal.

27. Three forces P, Q and R act on a body and keep it in equilibrium. The angle between P and Q is 135° and the angle between P and R is also 135°. Show that $P = Q\sqrt{2} = R\sqrt{2}$.

28. A weight of 12 N is suspended by two strings which pass over smooth pulleys and have weights of 8 N and 9 N attached to their free ends. Find the angle between the two strings.

29. Forces of 5 N, 4 N and 5 N respectively act along the sides AB, BC, CA of an equilateral triangle of side 2 m. Find the magnitude of their resultant, its direction, and the length of AX, where X is the point at which the line of action of the resultant cuts AB (or AB produced). (O. & C.*)

30. ABCD is a square of side 4 m. Forces 2 N, 4 N and 1 N act along AB, BC, CD respectively. Find the magnitude and direction (i.e. the angle made with AB) of the resultant.

Find where the line of action of the resultant cuts AB (or AB produced.) (O. & C.*)

CHAPTER 9

Friction

Experimental results

Consider a book lying on a horizontal table. If we apply a horizontal force H to the book, we may be able to push the book across the table. We notice that there are two possibilities. If the applied force H is fairly small, the book will stay at rest on the

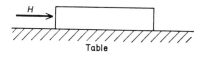

Table

Fig. 75

table. If the force H does not produce any acceleration of the book, there must be an equal and opposite force preventing acceleration. This force is called the *frictional force*, the force caused by *friction* between the book and the table. As H increases the frictional force increases so that it is still able to prevent motion, until it reaches its maximum value, after which the book slips. If we place the book on different surfaces we shall notice that the force required before it slips will not always be the same, and so we deduce that the greatest frictional force depends on the nature of the two surfaces in contact.

Experience suggests at first that friction always opposes motion, but closer observation shows that friction in fact opposes *relative* motion, so that a parcel can be placed in safety on the back of an electric truck, which then accelerates. Since the surface of the truck is rough, and the acceleration of the truck small, the parcel stays where it has been placed during the subsequent motion. The parcel does not move forwards or backwards in normal motion of the truck. The frictional force is sufficiently large to produce the small changes in the velocity of the parcel which are necessary for it to remain in position on the truck. Try placing the same parcel on a similar surface in a car. It will probably need to be fixed in position to prevent it sliding backwards when the car accelerates and forwards when the car brakes (you may have noticed this effect on parcels placed on the ledge of a car near the rear window).

In trying to prevent relative motion then, friction will often cause acceleration. Since the parcel stays in position on the truck while the truck is accelerating, the parcel is also accelerating. What force causes this acceleration? The only horizontal force acting on the parcel is the frictional force between the parcel and the truck, so that must be the force which is causing acceleration.

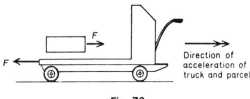

Fig. 76

Here we have an example of friction causing acceleration. The frictional force prevents the parcel slipping on the truck, that is, it opposes relative motion.

Consider also the driving wheels of a car. The engine turns these and, if the friction between the tyres and the road is sufficient, they cannot slip, and so the car moves forward. Think what would happen if you tried to start a car on an icy road! Again, when we walk, the friction between the ground and our shoes prevents us slipping. An athlete who wishes to accelerate rapidly wears shoes with spikes to prevent him slipping, as the frictional force produced by the ground is not adequate. Have you ever tried ice-skating?

What happens to the parcel on the truck when it is slipping? Experiments suggest that the maximum frictional force continues to act. This is not quite true, and for very high speeds and other exceptional circumstances there may be a considerable difference, but for our purpose this is adequate.

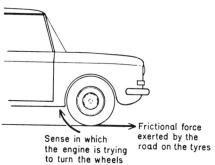

Sense in which
the engine is trying
to turn the wheels

Frictional force
exerted by the
road on the tyres

Fig. 77

Further experiments which we can perform show that the maximum frictional force exerted by a surface on a body is proportional to the normal reaction between the surface and the body, and NOT to the weight of the body, and also that the maximum frictional force is independent of the shape of the areas in contact. This last result is not used at this stage.

Laws of friction

Thus we can begin to formulate some laws of friction:

1. Friction always opposes relative motion.
2. The frictional force is just sufficient to prevent relative motion, up to a certain limiting maximum value. The maximum force is called limiting friction.
3. The limiting force is proportional to the normal reaction between the bodies in contact, the constant of proportion being called the coefficient of friction and usually denoted by μ.
4. The coefficient of friction depends only on the nature of the surfaces in contact, and not on the size or shape of the areas in contact.
5. When two bodies are moving relative to one another, the frictional force exerted by one body on the other will have its maximum value. These forces, are, of course, equal and opposite.

The first two laws can be expressed very clearly in algebra,

$$F \leqslant F_{max}$$

and Law 3 can be written

$$F_{max} = \mu R$$

giving

$$F \leqslant \mu R,$$

where F is the frictional force and R is the normal reaction. Of these laws, two are particularly important. The frictional force is LESS THAN OR EQUAL to a certain value, and the frictional force is proportional to the normal reaction, NOT to the weight of the body.

Example 1. *A body mass 3 kg is at rest on a horizontal table. The coefficient of friction between the body and the table is 0.5. A horizontal force F is applied to the body. Find the frictional force exerted by the table on the body if (i) F = 10 N, (ii) F = 20 N.*

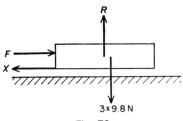

Fig. 78

Let the frictional force be X N. Since the weight of the body is 29.4 N, resolving vertically

$$R = 29.4 \text{ N}$$

Thus the maximum frictional force is 0.5 R, i.e. 14.7 N.

In (i), the force needed to prevent motion is only 10 N, so that maximum friction is not needed; so $X = 10$ and the frictional force is 10 N.

In (ii), however, the force necessary to prevent motion is 20 N. But the greatest force that friction can provide is 14.7 N, so that $X = 14.7$, and the frictional force is 14.7 N.

Example 2. *A body mass 5 kg is at rest on a rough horizontal plane: the coefficient of friction μ between the body and the floor is 0.6. A force of 20 N at 30° above the horizontal is applied to the body. Find the frictional force exerted by the floor on the body.*

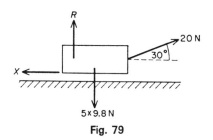

5 × 9.8 N

Fig. 79

Resolving vertically,

$$R + 20 \sin 30° = 49,$$
$$\Rightarrow R = 39$$

so the maximum frictional force is 0.6 × 39 N, i.e. 23.4 N. Since the body is in equilibrium, resolving horizontally,

$$X = 20 \cos 30° = 17.3.$$

This is less than the maximum frictional force, confirming that the body is in equilibrium and the frictional force is 17.3 N.

Notice that if the applied force had been 26 N instead of 20 N, the horizontal component would have been about 22.5 N. We might think that the body would still be at rest. But, resolving vertically, we now have

$$R + 26 \sin 30° = 49$$
$$\Rightarrow R = 36,$$

so that the maximum frictional force available is 0.6 × 36, i.e. 21.6 N, not enough to prevent motion.

Example 3. *A body of mass 2 kg is at rest on a rough horizontal surface; the coefficient of friction between the body and the surface* *is 0.4. A force* **F** *is applied to the body. Find the frictional force and describe the ensuing motion, if any, when*

(i) **F** = 2**i** + 4**k**,

(ii) **F** = 4**i** + 18**k**,

where **i** *is a horizontal unit vector and* **k** *is a vertical unit vector.*

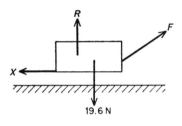

Fig. 80

For (i), using Fig. 80 and resolving vertically,

$$R + 4 = 19.6$$
$$R = 15.6.$$

The maximum frictional force is therefore 6.24 N which is greater than 2 N. Thus the frictional force is $-2\mathbf{i}$ N and the body stays at rest.

In (ii), resolving vertically,

$$R + 18 = 19.6$$
$$\Rightarrow R = 1.6$$

The maximum frictional force is now 0.64 N, not sufficient to prevent motion. Applying Newton's Second Law,

$$F = ma$$
$$\Rightarrow 4 - 0.64 = 2a$$
$$\Rightarrow a = 1.68.$$

therefore the frictional force is $-0.64\mathbf{i}$ N and the acceleration is 1.68 m s^{-2} horizontally.

Angle of friction

When a body is at rest on a rough table, the forces exerted by the table on the body are the normal reaction R and the frictional force F. These may be combined to give the force P, acting at an

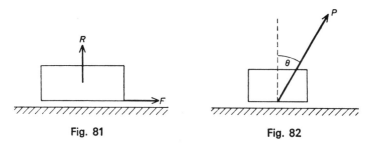

Fig. 81   Fig. 82

angle, say, θ to the normal. P and θ may be found by resolving horizontally and vertically:

$$P \sin \theta = F; P \cos \theta = R$$

$$\therefore \qquad P = \surd(F^2 + R^2), \qquad \tan \theta = F/R.$$

As F increases, θ increases. The maximum value of F is μR, so the greatest value of $\tan \theta = \mu$. This value of θ is called the angle of friction, and is usually denoted by λ, i.e.

$$\tan \lambda = \mu.$$

In the case when a body is at rest on a rough plane which can be tilted, if the angle of inclination of the plane to the horizontal is gradually increased, the body will eventually slide. The angle at which sliding takes place varies. One book placed on another book may slip when the angle of inclination of the lower book is 20°. A board-rubber slipped on my card table when the card-table was inclined at 40°.

Let us look at the forces acting, when slipping is just about to take place.

Let R denote the normal reaction of the table on the body,

$\quad$ F the frictional force acting on the body,

and mg, the weight of a body of mass m.

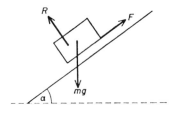

Fig. 83

Resolving along the plane,

$$mg \sin \alpha = F,$$

and perpendicular to the plane,

$$mg \cos \alpha = R.$$

Since sliding is about to take place the friction is limiting and $F = \mu R$. Therefore by division

$$\tan \alpha = \mu.$$

Thus the angle at which sliding takes place is equal to the angle of friction.

Use of the angle of friction

If we consider the total reaction between two bodies, instead of the normal reaction and the frictional force, this reduces by one the number of forces and we may be able to find an unknown force immediately by resolving in a convenient direction.

Example 1. *A body mass 5 kg is placed on a rough plane inclined at 30° to the horizontal. The angle of friction between the plane and the body is 20°. Find the force X, along the plane, necessary to prevent the body slipping down the plane.*

Since we want to find X only, we resolve perpendicular to the line of action of the reaction R between the body and the plane.

$$X \sin 70° = 5 \times 9.8 \times \sin 10°$$
$$\Rightarrow X = 9.05 \text{ N.}$$

If instead of applying X along the plane we had wished to

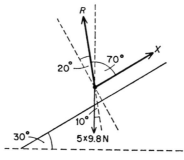

Fig. 84

prevent slipping by a horizontal force, we should have resolved in the same direction as before, and obtained, from Fig. 85,

$$X \sin 80° = 5 \times 9.8 \times \sin 10°$$
$$\Rightarrow X = 8.64 \text{ N}.$$

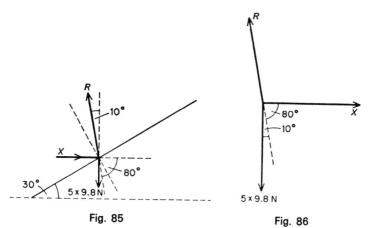

Fig. 85

Fig. 86

Fig. 85 shows the lines of action of the forces in relation to the plane and to the horizontal, but Fig. 86 shows the angles between the forces which we want to use and enables us to see clearly the directions in which it is best to resolve.

Least force required to move a body

Notice from these examples that the angle of inclination of the plane is 30° and the angle of friction 20°. The angle 10° in Figs. 85, 86 is the difference of these, as we would expect, since if the angle of inclination is equal to the angle of friction the body will rest in equilibrium, without any external force being necessary. If we wish to find the least force required to maintain equilibrium, the factor by which X is multiplied, sin 70° in the first example

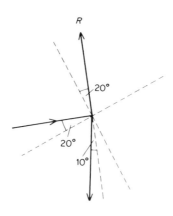

Fig. 87

and sin 80° in the second example, must be as large as possible. Sin 90° is the largest such factor and the least force necessary to maintain equilibrium is 8.51 N acting at 20° to the plane as in Fig. 87.

EXERCISE 15

In this exercise **i** and **j** denote unit vectors along horizontal axes Ox and Oy and **k** a unit vector along the vertical axis Oz.

1. A body mass 2 kg at rest on a rough horizontal surface is acted on by a horizontal force F. The coefficient of friction between the

body and the surface is 0.4. Find the frictional force and describe the motion of the body, if any, when

> (a) $F = 1$ N,
> (b) $F = 7$ N,
> (c) $F = 8$ N,
> (d) $F = 16$ N.

2. A lorry mass 2000 kg accelerates from rest at 0.5 m s^{-2}. What is the frictional force between the tyres of the lorry and the ground? In which direction does this frictional force act?

3. A body mass 30 kg is at rest on a rough horizontal floor: the coefficient of friction between the body and the floor is 0.5. A force 150i N acts on the body. Find
   (a) the frictional force on the body,
   (b) the acceleration of the body.

4. A body mass 3 kg is at rest on a rough horizontal table. The coefficient of friction between the body and the table is 0.4. Find the acceleration of the body if a horizontal force of 20 N acts on the body. What horizontal force acting on the body will produce an acceleration of 1 m s^{-2}?

5. A body of mass 10 kg is at rest on a rough horizontal table. Find the normal reaction X of the table on the body if
   (a) the weight is the only force which acts on the body,
   (b) a vertical force of 50 N upwards acts on the body,
   (c) a horizontal force of 20 N acts on the body,
   (d) a force of 30 N at 36° above the horizontal acts on the body,
   (e) a force of 30 N at 36° below the horizontal acts on the body.
   Find in each case the frictional force F which prevents motion. Find the greatest value of the ratio $F : X$. Comment on the value of the coefficient of friction between the body and the surface.

6. A body mass 10 kg is at rest on a rough horizontal floor: the coefficient of friction between the body and the floor is 0.4. A force (70i − 30k) N acts on the body. Find
   (a) the normal reaction between the body and the surface,
   (b) the frictional force on the body,
   (c) the acceleration of the body.

7. A wooden block, mass 100 kg is being dragged by two men A and B over rough ground; the coefficient of friction between the block and the ground is 0.9. A's rope is inclined at 15° to the direction in which the block moves. A exerts a pull of 400 N, and the block moves with constant velocity. Find
   (a) the magnitude of the force exerted by B,
   (b) the direction in which B pulls.

8. Two children Clare and Elizabeth are pulling a toboggan mass 20 kg; the coefficient of friction between the toboggan and the snow is 0.25. If the toboggan moves with constant velocity $2i$ m s^{-1} and the pull of Clare is $(40i + 10j)$ N, find the vector which describes the pull of Elizabeth.

9. Two men Bill and Bert are pulling a wooden block mass 100 kg over a concrete floor. The coefficient of friction between the block and the floor is 0.7. Bill exerts a force of $(400i + 400j)$ N and Bert a force of $(300i - 400j)$ N. Find in vector form,
(a) the frictional force on the block,
(b) the acceleration of the block.

10. A parcel mass 2 kg is initially at rest on an electric truck; the coefficient of friction between the parcel and the truck is 0.8. Find the frictional force of the truck on the parcel if
(a) the truck accelerates at 1 m s^{-2},
(b) the truck accelerates at 10 m s^{-2}.

11. A uniform rod length 2 m rests on rough horizontal ground and leans against a smooth vertical wall. The mass of the rod is 20 kg. If the rod is inclined to the horizontal at 45°, calculate the frictional force at the foot of the rod. What can you say about the coefficient of friction between the rod and the ground?

12. A rod AB length 2 m mass 40 kg rests with one end A on rough horizontal ground at an angle of 60° with the horizontal. The rod is supported by a smooth bar placed at a point 1.4 m from A. Find the frictional force which the ground exerts on the rod. What is the least value of the coefficient of friction between the ground and the rod?

13. A light ladder 4 mm long is inclined at 45° to the horizontal, resting against a cube edge 3 mm, mass 1 g. A Lilliputian, mass 0.2g, is placed gently by Gulliver halfway up the ladder. Another Lilliputian has his foot on the bottom of the ladder to prevent slipping. Denoting the coefficient of friction between the cube and the ground by μ, what happens to the ladder (and the first Lilliputian!) if the cube is resting on ice ($\mu = 0.05$)? What happens if the cube is resting on concrete ($\mu = 0.8$) and the Lilliputian tries to climb carefully to the top of the ladder?

14. A body mass 30 kg is at rest on a rough plane inclined at 30° to the horizontal. Find
(a) the normal reaction between the plane and the body,
(b) the frictional force which acts on the body if it is just sufficient to prevent the body slipping down the plane,
(c) the angle of friction between the body and the plane if the body is in limiting equilibrium.

(d) the least force acting on the body which will cause it to move up the plane.

15. A body mass 3 kg is at rest on a rough plane inclined at 30° to the horizontal. The coefficient of friction between the body and the plane is 0.2. Find

(a) the additional force along the plane which is necessary to prevent the body slipping down the plane,
(b) the force along the plane which is necessary to cause the body to move up the plane with constant velocity,
(c) the force along the plane which will cause the body to move up the plane with an acceleration of 1 m s^{-2}.

16. A boat of mass 100 kg can just be pulled down a slipway inclined at 30° to the horizontal by a force of 150 N acting down the slipway. Find the coefficient of friction between the boat and the slipway. Find also the least horizontal force which is sufficient to pull the boat down the slipway.

17. The truck in question 10 has stopped on a hill which makes an angle of 30° with the horizontal. Describe what happens to the parcel if

(a) the truck accelerates up the hill at 1 m s^{-2},
(b) the truck accelerates up the hill at 10 m s^{-2},
(c) the truck accelerates downhill at 1 m s^{-2},
(d) the truck accelerates downhill at 10 m s^{-2}.

18. A body mass 5 kg is at rest on a rough plane inclined at 20° to the horizontal. The angle of friction between the body and the plane is 20°. Find the least force necessary to cause the body to move up the plane with constant velocity.

19. The plane in question 18 is now inclined at 25° to the horizontal. Find the least force necessary to prevent the body slipping down the plane.

20. A chain AB length 5 m, mass 10 kg is in equilibrium with part of itself in a straight line on a rough horizontal table while the rest hangs vertically over the edge of the table. If the coefficient of friction between the chain and the table is 0.8, find the greatest length of chain which can hang vertically.

If a horizontal force of 20 N acting away from the edge of the table, is now applied at A, the end of the chain on the table, what length of chain can hang vertically?

If instead a force of 20 N vertically downwards is applied at B, what length of chain can hang freely now?

21. A body mass 4 kg is at rest on a rough horizontal plane, the angle of friction between the body and the plane is 20°. Find the magnitude and direction of the least force necessary to move the body along the plane.

22. A body mass 1 kg is placed on a rough plane inclined at 40° to the horizontal. The angle of friction between the body and the plane is 30°. Find the force along the plane needed to prevent the body slipping down the plane.

23. With the data of question 22, find the magnitude and direction of the least force necessary to prevent the body slipping down the plane.

24. With the data of question 22, find the force along the plane which is just able to move the body up the plane.

CHAPTER 10

Centre of Gravity

Symmetrical bodies

We have referred earlier to the centre of gravity of simple bodies, by which we mean the point at which the weight of the body may be supposed to act, and we have relied on our experience of simple experiments. The nearest we can get to a mathematical 'rod' is probably an unsharpened pencil, and we know that can be supported by a single force acting at a point M, which by symmetry must be at the midpoint of the pencil. Since the pencil is in equilibrium, the downward forces (the weights of the parts of the pencil) must be equal to the upward forces (the force exerted by the support on the pencil at M).

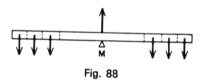

Fig. 88

So the force exerted by gravity on each part of the pencil, that is, the weights, can be replaced by a single force acting at M. We call M the centre of gravity or sometimes the centre of mass, of the pencil.

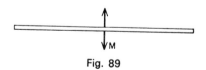

Fig. 89

If the pencil had been sharpened, the point at which the pencil could be supported would not be at the middle point of the pencil, and so the centre of gravity would not be at M.

Similarly, experience shows that we can balance a book by applying a single force at one point. If the book is uniform, then by symmetry the point of support must be at the intersection of the axes of symmetry.

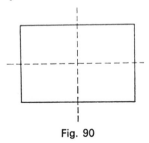

Fig. 90

Non-symmetrical bodies

But what happens when the body is not symmetrical? There are many situations in which it is convenient and correct to take the weight of a body as acting at one point, the centre of gravity, and we need to investigate how to find the position of the centre of gravity of such bodies.

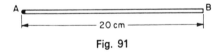

Fig. 91

Suppose that to one end A of a rod AB length 20 cm mass 2 kg, a small mass 1 kg regarded as a particle, is fixed, as in Fig. 92. Where is the centre of gravity of this system? The weight of the rod is $2g$ N and the weight of the particle is $1g$ N, so the weight

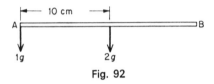

Fig. 92

of the system is $3g$ N. But if $2g$ N at M and $1g$ N at B are to be replaced completely by a single force of $3g$ N, the moments about any axis of the two systems must be equal. In particular, the moment of the $3g$ N about a horizontal axis through A perpendicular to AB must equal the sum of the moments of the $2g$ N and $1g$ N about the same axis.

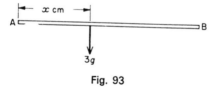

Fig. 93

If the centre of gravity is x cm from A, then taking moments about A we have

$$2g(10) = 3g(x)$$
$$\Rightarrow x = 6\tfrac{2}{3}$$

i.e. the centre of gravity is $6\tfrac{2}{3}$ cm from A.

Note that the position of the centre of gravity divides the distance between the lines of action of the forces in the ratio 1 : 2, the inverse ratio of the magnitudes of the forces.

Any symmetry that the body possesses should always be used at the beginning of a problem to help to locate the centre of gravity. This may reduce a two- or three-dimensional problem even to a one-dimensional problem.

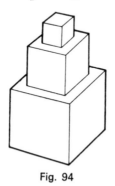

Fig. 94

Suppose a child places a uniform cubical brick, edge 2 cm, on the middle of a similar brick, edge 3 cm, then a third brick edge 1 cm on top of the first, as in Fig. 94. We wish to find the position of the centre of gravity of the solid so formed. From the symmetry of each brick, the centre of gravity of a brick is at the centre of the brick. Since the bricks are similar, their masses are proportional to the cube of the lengths of the edges, and so may be taken to be m, $8m$, $27m$. The problem has reduced at once to finding the centre of mass of masses m, $8m$ and $27m$ placed in a straight line.

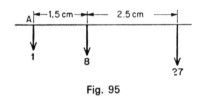

Fig. 95

Calling the centre of the smallest cube A and taking moments about a horizontal axis through A perpendicular to the axis of symmetry,

$$8mg(1.5) + 27mg(4) = 36mg(x)$$
$$\Rightarrow x = \tfrac{10}{3}.$$

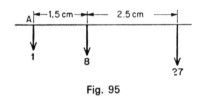

Fig. 96

Thus the centre of gravity of the three cubes is $3\tfrac{1}{3}$ cm from A, i.e. $2\tfrac{4}{5}$ cm above the base of the large cube. Checking this with the positions of the centres of gravity of the three separate bodies, it is about three-quarters the height of the largest cube, which is reasonable.

Use of coordinate systems

If the forces are not applied at points in a straight line, it is necessary to work in two dimensions, and later on of course we shall have to use three dimensions. It is usually easy to describe the positions of the points of application of the forces and the position of the centre of gravity by the notation of Cartesian coordinate geometry. Choose the coordinate axes as simply as is possible, so that they pass through as many points of application as possible, and then take moments about these axes.

Example 1. *Masses 1 kg, 2 kg, 3 kg and 4 kg are placed one at each corner A, B, C and D of a square table, side 1 m. Find the position of the centre of gravity of the masses.*

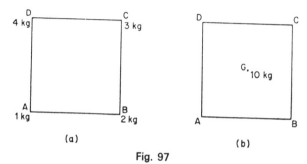

Fig. 97

Denote the coordinates of the centre of gravity G by $(\bar{x}, \bar{y})$, and take moments about AB.

$$4g(1) + 3g(1) = 10g(\bar{y})$$
$$\Rightarrow \bar{y} = 0.7.$$

Taking moments about AD,

$$2g(1) + 3g(1) = 10g(\bar{x})$$
$$\Rightarrow \bar{x} = 0.5,$$

i.e. the coordinates of G are (0.5, 0.7) referred to AB and AD as coordinate axes, or the centre of gravity is 0.5 m from AD and 0.7 m from AB.

Notice how easily this could be expressed in vector notation. The position vectors are:

$$0\mathbf{i} + 0\mathbf{j} \text{ for the 1 kg mass,}$$
$$1\mathbf{i} + 0\mathbf{j} \text{ for the 2 kg mass,}$$
$$1\mathbf{i} + 1\mathbf{j} \text{ for the 3 kg mass,}$$
$$0\mathbf{i} + 1\mathbf{j} \text{ for the 4 kg mass.}$$

The expression

$$1g(0\mathbf{i} + 0\mathbf{j}) + 2g(\mathbf{i} + 0\mathbf{j}) + 3g(\mathbf{i} + \mathbf{j}) + 4g(0\mathbf{i} + \mathbf{j})$$

simplifies to

$$(2g + 3g)\mathbf{i} + (3g + 4g)\mathbf{j}.$$

Here the $\mathbf{i}$ vectors give the moments about the y-axis and the $\mathbf{j}$ vectors the moments about the x-axis. Considering the final system, $10g(\bar{x}\mathbf{i} + \bar{y}\mathbf{j})$ also gives the moments about the two coordinate axes. Thus

$$5g\mathbf{i} + 7g\mathbf{j} = 10g(\bar{x}\mathbf{i} + \bar{y}\mathbf{j}),$$

from which $\qquad 5g = 10g\bar{x}, \qquad 7g = 10g\bar{y},$

i.e. $\qquad\qquad \bar{x} = 0.5 \quad \text{and} \quad \bar{y} = 0.7$

as before.

More generally, if masses $m_1, m_2, m_3, \ldots$ are placed at points

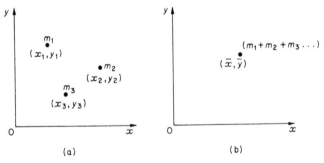

Fig. 98

whose coordinates are $(x_1, y_1), (x_2, y_2), (x_3, y_3), \ldots$, moments about Ox give

$$m_1y_1 + m_2y_2 + m_3y_3 \ldots = (m_1 + m_2 + m_3 \ldots)\bar{y}$$

and moments about Oy give

$$m_1x_1 + m_2x_2 + m_3x_3 \ldots = (m_1 + m_2 + m_3 \ldots)\bar{x}$$

i.e.
$$\bar{x} = \frac{m_1 x_1 + m_2 x_2 + m_3 x_3 \ldots}{m_1 + m_2 + m_3 \ldots},$$

and
$$\bar{y} = \frac{m_1 y_1 + m_2 y_2 + m_3 y_3 \ldots}{m_1 + m_2 + m_3 \ldots},$$

as illustrated in the following example.

Example 2. *Find the centre of gravity of masses 3m at (4,1), 2m at (−1, −2) and 4 m at (1, −2).*

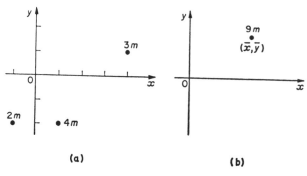

Fig. 99

Moments about Ox for both systems are equal, i.e.

$$3mg(1) + 2mg(-2) + 4mg(-2) = 9mg(\bar{y}),$$
$$9\bar{y} = -9,$$
$$\bar{y} = -1,$$

and moments about Oy must be equal,

$$3mg(4) + 2mg(-1) + 4mg(1) = 9mg(\bar{x}),$$
$$9\bar{x} = 14,$$
$$\bar{x} = \tfrac{14}{9}.$$

The coordinates of the centre of gravity are therefore $(\tfrac{14}{9}, -1)$.

Once again, notice how this could be expressed in terms of vectors. We wish to find the position vector of the centre of gravity of masses $3m$ at $(4\mathbf{i} + \mathbf{j})$, $2m$ at $(-\mathbf{i} + -2\mathbf{j})$ and $4m$ at $(\mathbf{i} - 2\mathbf{j})$. Taking moments,

$$3mg(4\mathbf{i} + \mathbf{j}) + 2mg(-\mathbf{i} - 2\mathbf{j}) + 4mg(\mathbf{i} - 2\mathbf{j}) = 9mg(\bar{x}\mathbf{i} + \bar{y}\mathbf{j})$$
$$\Rightarrow 14\mathbf{i} - 9\mathbf{j} = 9\bar{x}\mathbf{i} + 9\bar{y}\mathbf{j},$$
$$\Rightarrow \bar{x} = \tfrac{14}{9}, \qquad \bar{y} = -1.$$

The position vector of the centre of gravity is therefore $\tfrac{14}{9}\mathbf{i} - \mathbf{j}$.

Centre of gravity of a lamina

We considered earlier how to find the position of the centre of gravity of a pencil and a book by symmetry. This can be extended to the circle and a few other shapes, but obviously only a few bodies admit solution by this method. Many others, however may be divided into rods, rectangular laminae,* circles, discs, etc.

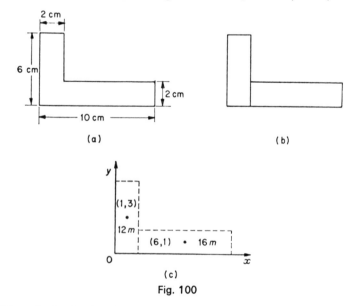

(a)   (b)

(c)

Fig. 100

Thus the lamina in Fig. 100(a) may be divided into two rectangles as in Fig. 100(b) and the weight of each may be presumed to act

* A lamina is a plane sheet whose thickness may be neglected.

at the centre of symmetry of each part. As the lamina is uniform, the masses, and hence the weights, are proportional to the surface areas. Choosing coordinate axes as in Fig. 100(c), the problem reduces to finding the position of the centre of gravity of $12m$ at $(1,3)$ and $16m$ at $(6,1)$. Taking moments about Ox,

$$12mg(3) + 16mg(1) = 28mg(\bar{y}),$$
$$\Rightarrow \bar{y} = \tfrac{13}{7}.$$

Taking moments about Oy,

$$12mg(1) + 16mg(6) = 28mg(\bar{x}),$$
$$\Rightarrow \bar{x} = \tfrac{27}{7},$$

i.e. the coordinates of the centre of gravity are $(\tfrac{27}{7}, \tfrac{13}{7})$.

Again, this could have been worked in vector notation. $12m$ at the point whose position vector is $(\mathbf{i} + 3\mathbf{j})$ and $16\ m$ at the point, position vector $(6\mathbf{i} + \mathbf{j})$ give

$$12mg(\mathbf{i} + 3\mathbf{j}) + 16mg(6\mathbf{i} + \mathbf{j}) = 28mg(\bar{x}\mathbf{i} + \bar{y}\mathbf{j}),$$
$$\Rightarrow 108\mathbf{i} + 52\mathbf{j} = 28(\bar{x}\mathbf{i} + \bar{y}\mathbf{j}).$$

The position vector of the centre of gravity is $\tfrac{27}{7}\mathbf{i} + \tfrac{13}{7}\mathbf{j}$.

Centre of gravity of a triangular lamina

The method of division into rods enables us to find the position of the centre of gravity of a triangular lamina. If we divide the lamina into thin rods, as in Fig. 101, the centre of gravity of each

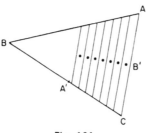

Fig. 101

rod must be at the midpoint of the rod. Thus the centre of gravity of the triangular lamina must be on the line through the midpoint of each rod. These midpoints lie on a straight line BB', one of the medians of the triangle. Similarly, division into rods parallel to BC shows that the centre of gravity of the lamina must lie on the median AA'. It was shown in Chapter 6 that the medians of a triangle concur at a point of trisection of each median, so that the centre of gravity of a triangular lamina is at the point of intersection of the medians, one third of the way up a median.

Bodies with parts cut away

The method of equating the moments of the two systems to enable us to find the centre of gravity can also be used if a body had a part removed. Consider a circular lamina from which a smaller circular lamina has been removed, as in Fig. 102. By symmetry,

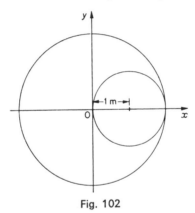

Fig. 102

the centre of gravity of the remaining lamina lies on Ox. If the mass of the original lamina was $4m$, the mass of the part cut away is $1m$. The moment of the original lamina minus the moment of the part removed must equal the moment of the remaining lamina. Since the centre of the original lamina has been taken as origin, the moment of the original lamina is zero. Therefore,

$$4mg(0) - mg(1) = 3mg(\bar{x}),$$
$$\Rightarrow \bar{x} = -\tfrac{1}{3}.$$

The centre of gravity is $\tfrac{1}{3}$ cm to the **LEFT** of the centre of the original circle. Strange as this often seems, it is reasonable once we have realised that the centre of gravity of the original circle was at O, by symmetry, and some part was removed to the right of O. There must therefore be less matter, and hence less weight to the right of O than to the left, so the centre of gravity must be left of O.

Hanging bodies

At the beginning of this chapter we noticed that the weight of a pencil acted through the same point as the reaction at the support, since these were the only two forces acting on the pencil. This often enables us to find the manner in which irregular objects will hang under the action of their own weights.

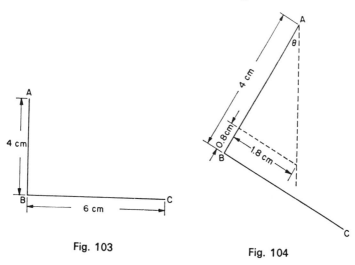

Fig. 103   Fig. 104

Consider a uniform wire, ABC, 10 cm long bent as in Fig. 103 and suspended from A. Taking moments about AB and BC we find that the centre of gravity G of the wire is 1.8 cm from AB and 0.8 cm from BC. The weight of the wire can therefore be replaced by a single vertical force $10mg$ at G. Since the reaction at A is the only other force acting on the wire, these two must act in the same straight line. Considering the trigonometry of Fig. 104, we see that the angle θ which AB makes with the vertical is such that

$$\tan \theta = \frac{1.8}{3.2}$$

i.e. $\theta \simeq 29\frac{1}{2}°$.

EXERCISE 16

1. Find by symmetry the position of the centre of gravity of each of the following,
   (a) lamina bounded by two concentric circles,
   (b) a ring,
   (c) a ball as used in Rugby football,
   (d) a prism whose cross-section is an equilateral triangle.
2. To a light uniform rod AB 2 m long are fixed masses 3 kg at A and 2 kg at B. Find the distance from A of the centre of gravity of the system.
3. To a light uniform rod AB 2 m long are fixed masses 3 kg at A, 2 kg at B and 5 kg at the mid point of the rod. Find the distance from A of the centre of gravity of the system.
4. A rod AB mass 2 kg length 0.9 m is rigidly joined to another rod BC mass 4 kg length 0.6 m so that ABC is a straight line. Find the distance from A of the centre of gravity of the system.
5. To a uniform rod AB mass 2 kg length 0·4 m are fixed masses 3 kg at A and 5 kg at B. Find the distance from A of the centre of gravity of the system.
6. To a uniform rod AB mass 3 kg length 1.4 m are fixed masses 6 kg at A and 5 kg at B. Find the distance from A of the centre of gravity of the system.
7. To a uniform rod AB mass 2 kg, length 0.4 m is fixed a uniform rod BC mass 3 kg, length 0.6 m so that angle ABC = 90°. Find the distance from AB and from BC of the centre of gravity of the system.
8. If the rods in Q.7 are fixed so that angle ABC = 60°, find the distance of the centre of gravity of the system from AB and from the line through B perpendicular to AB.
9. Find the distance from AB and from BC of the centre of gravity of the uniform lamina illustrated in Fig. 105.
10. If this lamina in Fig. 105 is freely suspended from A, find the angle AB will make with the vertical.
11. Find the distance from AB and from BC of the centre of gravity of the uniform lamina illustrated in Fig. 106.

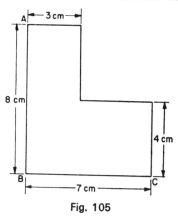

Fig. 105

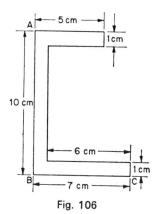

Fig. 106

12. If the lamina in Fig. 106 is freely suspended from A, find the angle AB will make with the vertical.

13. In the uniform rectangular plate ABCD, AB = 8 cm, BC = 4 cm. A circular disc radius 1 cm centre 5 cm from BC and 1.5 cm from AB is cut away. Find the distance from BC and from AB of the centre of gravity of the remaining body.

14. Find the coordinates of the centre of gravity of a mass 4 kg at (2, −3), a mass 3 kg at (−3, 5) and a mass 1 kg at (1, −3).

15. Find the coordinates of the centre of gravity of masses 2 kg at (1, −1), 3 kg at (−3, 5) and 5 kg at (−2, 1).

16. Find the position vector of the centre of gravity of masses 1 kg at P_1, 4 kg at P_2 and 3 kg at P_3, when the position vectors of P_1, P_2 and P_3 are $(i + j)$, $(3i + 2j)$ and $(−3i − 3j)$ respectively.

17. Find the position vector of the centre of gravity of masses 5 kg at P_1, 2 kg at P_2 and 3 kg at P_3, when the position vectors of P_1, P_2 and P_3 are $(−i −j)$, $(3i + 2j)$ and $(−3i − 3j)$ respectively.

18. Find the position vector of the centre of gravity of a mass m at ai and a mass M at $2ai + bj$.

19. A non-uniform rod AB length 2 m is suspended by two vertical strings attached one to each end of the rod. The tension in the string at A is 14 N and in the string at B 26 N. Find the distance of the centre of gravity of the rod from A.

20. A narrow straight hole is bored through the centre of each of three solid metal spheres, radii 1 cm, 2 cm and 4 cm. The spheres are threaded onto a metal rod, of the same material as the spheres, which just fills the holes in the spheres. The largest sphere touches

each of the smaller ones. Find the distance of the centre of gravity of the system from the centre of the largest sphere.

21. Find the distances from AB and BC of the centre of gravity of the lamina illustrated in Fig. 107 in which angles DAB and ABC are right angles.

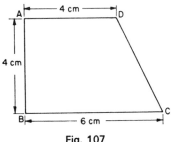

Fig. 107

22. By dividing the figure into two triangles find the distance from AB of the centre of gravity of the lamina illustrated in Fig. 108, in which AB is parallel to DC.

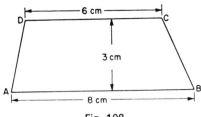

Fig. 108

23. If the three spheres in question 20 are hollow, so that the mass of each sphere is proportional to its surface area, and are threaded onto a wire of negligible mass in the same manner as before, find the distance of the centre of gravity of the system from the centre of the large sphere.

24. From a uniform circular metal disc, centre O and radius 10 cm, are cut two circular discs centres A and B, radii 1 cm and 2 cm respectively. The length of OA is 4 cm, of OB is 5 cm, and angle AOB is a right angle. Find the position of the centre of gravity of the remaining body, referred to OA and OB as axes. Find also the angle OA makes with the vertical when the disc is suspended freely from O.

Use of calculus

Applications to laminae

To find the centre of gravity of more difficult laminae, we can use the methods of calculus. The area of the lamina is divided into a number of elements, and the sum of the moments of these elements about any convenient axis is equal to the moment of the whole about that axis. Suppose we wish to find the centre of gravity of the area between the curve $y = f(x)$, the x-axis and the ordinates $x = a$ and $x = b$.

The area is divided into elements parallel to the y-axis. The area of the element in Fig. 109 lies between $y \, \delta x$ and $(y + \delta y) \, \delta x$,

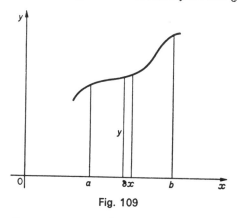

Fig. 109

and its mass lies between $my \, \delta x$ and $m(y + \delta y)\delta x$, where m is the mass per unit area. If A is the area under the curve, taking moments about the y-axis,

$$\Sigma mxy \, \delta x < mA\bar{x} < \Sigma mx(y + \delta y)\delta x.$$

In the limit, therefore,

$$A\bar{x} = \int_a^b xy \, \mathrm{d}x$$

$$\Rightarrow \bar{x} = \frac{1}{A} \int_a^b xy \, \mathrm{d}x,$$

The distance of the centre of gravity of the element from the x-axis is approximately $\frac{1}{2}y$ and so

$$A\bar{y} = \int_a^b y \, dx \left(\tfrac{1}{2}y\right)$$

$$\Rightarrow \bar{y} = \frac{1}{A} \int_a^b \tfrac{1}{2}y^2 \, dx.$$

As always, look for any symmetry which the body has to see if that will enable us to write down one or more coordinates without any detailed calculation.

Example. *Find the coordinates of the centre of gravity of the area included between the x-axis and the curve* $y = x^2 - 3x + 2$.

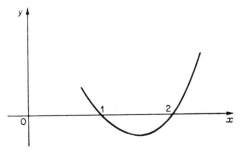

Fig. 110

The curve cuts the x-axis where $x = 1$ and $x = 2$. Recognizing that the curve is part of a parabola with its axis parallel to the y-axis, we know that it is symmetrical about $x = \frac{3}{2}$ and can write down

$$\bar{x} = \tfrac{3}{2}.$$

The area defined is given by

$$A = \int_1^2 y \, dx = \int_1^2 (x^2 - 3x + 2) \, dx$$

$$= \left[\frac{x^3}{3} - \frac{3x^2}{2} + 2x\right]_1^2$$

$$= -\tfrac{1}{6}.$$

Thus taking moments about the x-axis,

$$A\bar{y} = \tfrac{1}{2}\int_1^2 y^2\,dx$$

$$= \tfrac{1}{2}\int_1^2 (x^4 - 6x^3 + 13x^2 - 12x + 4)dx$$

$$= \tfrac{1}{2}\left[\tfrac{1}{5}x^5 - \tfrac{3}{2}x^4 + \tfrac{13}{3}x^3 - 6x^2 + 4x\right]_1^2$$

$$= \tfrac{1}{60}$$

Since $A = -\tfrac{1}{6}$, therefore $\bar{y} = -\tfrac{1}{10}$. Hence the centre of gravity is at $(\tfrac{3}{2},\, -\tfrac{1}{10})$.

Example 2. *Find the coordinates of the centre of gravity of the area in the first quarter bounded by the parabola $y^2 = 4x$, the x-axis and the ordinate $x = 1$.*

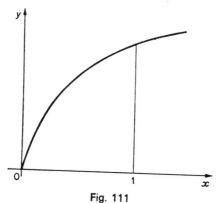

Fig. 111

The area defined is given by

$$A = \int_0^1 y\,dx$$

$$= \int_0^1 2x^{1/2}\,dx$$

$$= \left[\tfrac{4}{3}x^{3/2}\right]_0^1 = \tfrac{4}{3}.$$

Taking moments about the y-axis,

$$A\bar{x} = \int_0^1 xy \, \mathrm{d}x$$

$$\Rightarrow \tfrac{4}{3}\bar{x} = \int_0^1 2x^{3/2} \, \mathrm{d}x$$

$$= \left[\tfrac{4}{5}x^{5/2} \right]_0^1$$

$$= \tfrac{4}{5}$$

$$\Rightarrow \bar{x} = \tfrac{3}{5}.$$

Taking moments about the x-axis,

$$A\bar{y} = \tfrac{1}{2}\int_0^1 y^2 \, \mathrm{d}x$$

$$\Rightarrow \tfrac{4}{3}\bar{y} = \int_0^1 2x \, \mathrm{d}x$$

$$= \left[x^2 \right]_0^1$$

$$= 1$$

$$\Rightarrow \bar{y} = \tfrac{3}{4}.$$

The centre of gravity therefore is at $(\tfrac{3}{5}, \tfrac{3}{4})$.

Applications to solids of revolution

When a solid can be formed by rotating a lamina about a suitable axis, we can usually find one coordinate of the centre of gravity by symmetry. To find the other coordinate it may be convenient to divide the solid into elements perpendicular to the axis of rotation.

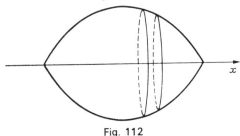

Fig. 112

If the axis of rotation is the x-axis, as in Fig. 112, the volume of the element will lie between $\pi y^2 \delta x$ and $\pi(y + \delta y)^2 \delta x$. If the volume of the solid is V, taking moments about the y-axis,

$$\Sigma \pi x y^2 \, \delta x < \bar{x} V < \Sigma \pi(x + \delta x)(y + \delta y)^2 \, \delta x$$

$$\Rightarrow \bar{x} V = \pi \int x y^2 \, \mathrm{d}x$$

$$\Rightarrow \bar{x} = \frac{\pi}{V} \int x y^2 \, \mathrm{d}x.$$

$\bar{y} = 0$, of course, by symmetry.

Example 1. *A solid is formed by rotating about the x-axis the area bounded by $y = x^2$, $y = 0$ and $x = 2$. Find the coordinates of the centre of gravity of this solid.*

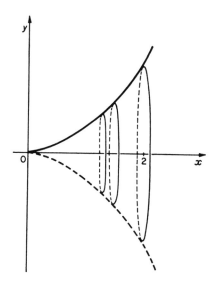

Fig. 113

By symmetry, $\bar{y} = 0$. Dividing the solid into discs the volume V is given by

$$V = \int_0^2 \pi y^2 \, \mathrm{d}x$$

$$= \pi \int_0^2 x^4 \, \mathrm{d}x$$

$$= \pi \left[\tfrac{1}{5} x^5 \right]_0^2$$

$$= \tfrac{32}{5} \pi.$$

Taking moments about the y-axis,

$$V\bar{x} = \int_0^2 x \pi y^2 \, \mathrm{d}x$$

$$\Rightarrow \tfrac{32}{5} \pi \bar{x} = \pi \int_0^2 x^5 \, \mathrm{d}x$$

$$= \pi \left[\tfrac{1}{6} x^6 \right]_0^2$$

$$= \tfrac{32}{3} \pi$$

$$\Rightarrow \bar{x} = \tfrac{5}{3}.$$

The centre of gravity is at $(\tfrac{5}{3}, 0)$.

Notice that $\bar{x}$ is nearer 2 than 0, since there is considerably more of the solid to the right of $x = 1$ than to the left.

Example 2. *A solid is formed by rotating about the x-axis the area bounded by* $y = \cos x$ *between* $x = \tfrac{1}{2}\pi$ *and* $x = \tfrac{3}{2}\pi$. *Find the centre of gravity of this solid.*

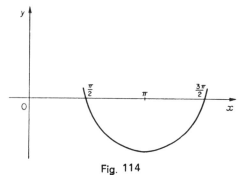

Fig. 114

Since the solid is obtained by rotation about $y = 0$, $\bar{y} = 0$.
Since the cosine curve is symmetrical about $x = \pi$, $\bar{x} = \pi$.
Hence the centre of gravity is at $(\pi, 0)$.

In these simple examples, we have considered only bodies of uniform density, and we have taken the mass per unit area or per unit volume as our unit of density. In problems in which the density varies, it is necessary to express this variation algebraically, and then to proceed as in the above examples. (See questions 22, 23, 24 in Exercise 17.)

It is most desirable that the principle of division into elements and taking of moments should be understood, and the reader is urged not to rely on the formulae which can be devised.

EXERCISE 17

1. Find the centre of gravity of the uniform lamina bounded by the curve $y = x^2$, the x-axis and the ordinate $x = 1$.
2. Find the centre of gravity of the uniform lamina bounded by the curve $y = x^3$, the x-axis and the ordinates $x = 1$ and $x = 2$.
3. Find the centre of gravity of the uniform lamina bounded by the curve $y = x^2 - 1$ and the x-axis.
4. Find the centre of gravity of the uniform lamina bounded by the curve $y = 2x - x^2$ and the x-axis.
5. Find the centre of gravity of the uniform lamina bounded by the curve $y = x^2$, the y-axis and the line $y = 1$.
6. Find the centre of gravity of the uniform triangular lamina bounded by the lines $3y = 4x$, $y = 0$ and $x = 3$.
7. Find the centre of gravity of the uniform lamina bounded by the curve $y = x^2 - 4x$, the x-axis and the lines $x = 1$ and $x = 2$.
8. Find the centre of gravity of the uniform lamina bounded by the curve $y = x^2$ and the line $y = x$.
9. Find the centre of gravity of the uniform lamina bounded by the curve $y^2 = 4x$ and the line $y = x$.
10. Find the centre of gravity of the uniform bounded by the curves $y = \sin x$ and $y = -\sin x$ between $x = 0$ and $x = \pi$.
11. Find the centre of gravity of the uniform lamina bounded by $y = \sin x$ and the x-axis, between $x = -\pi$ and $x = \pi$.

12. Find the centre of gravity of the solid formed when the area bounded by $y = x^2$, $x = 1$ and the x-axis is rotated completely about the x-axis.

13. Find the centre of gravity of the solid formed when the area bounded by $y = x^3$, $x = 1$ and the x-axis is rotated completely about the x-axis.

14. Find the centre of gravity of the solid formed when the area bounded by $y = x^{\frac{1}{2}}$, $x = 1$ and the x-axis is rotated completely about the x-axis.

15. Find the centre of gravity of the solid formed when the area bounded by $y = 3x^2$, $x = 1$ and $x = 2$ is rotated completely about the x-axis.

16. Find the centre of gravity of the solid formed when the area bounded by $y = x^2$, $y = 1$ and the y-axis is rotated completely about the y-axis.

17. Write down by symmetry the coordinates of the centre of gravity of the solid formed when the area bounded by $y = x^2 - x$ and the x-axis is rotated completely about the x-axis.

18. Find the centre of gravity of the solid formed when the area bounded by $y = 4x - x^2$ and the x-axis is rotated completely about the x-axis.

19. Find the centre of gravity of the solid formed when the area bounded by $y^2 - 4y = x$ and the y-axis is rotated completely about the y-axis.

20. Find the centre of gravity of the solid formed when the area bounded by the x-axis and $y = \sin x$ between $x = 0$ and $x = \pi$ is rotated completely about the x-axis.

21. Find the centre of gravity of the solid formed when the area bounded by the x-axis and $y = \sin 2x$ between $x = 0$ and $x = \frac{1}{2}\pi$ is rotated completely about the x-axis.

22. Find the centre of gravity of a rod AB length $2a$, if the density at any point in the rod is proportional to the distance of that point from A.

23. Find the centre of gravity of the lamina bounded by $y^2 = x^4$ and $x = 1$, if the density at any point in the lamina is proportional to the distance of that point from the y-axis.

24. Find the centre of gravity of the solid formed when the lamina in question 23 is rotated completely about the x-axis.

25*. Find the centre of gravity of a length of uniform wire bent in the arc of a semi-circle radius 5 cm.

* requires $\int \cos \theta \, d\theta = \sin \theta + c.$

CHAPTER 11

Momentum and Impulse

When we were considering Newton's Second Law, in Chapter 4, we saw that if we know the force acting on a body of given mass we can find the acceleration of the body. Knowing also the velocity at the beginning of an interval of time we can find the final velocity. Thus a body mass 4 kg acted on by a force 16 N will have an acceleration of $\frac{16}{4}$, i.e. 4 m s^{-2}.

If the velocity of the body is initially 2 m s^{-1} in the same direction as the force, then having had an acceleration of 4 m s^{-2} for 6 s the velocity of the body will have increased to 26 m s^{-1}. In this way we can find the velocity of the body at any moment we wish.

It may well be that we are not interested in the acceleration and wish only to find the final velocity. Can we do so without first finding the acceleration? Let us express the problem algebraically.

A body of mass m under the action of a constant force F will have an acceleration F/m. If the initial velocity is u, the final velocity v after time t is given by

$$v = u + \frac{F}{m} t,$$

again assuming u and F are in the same direction.
From this equation we obtain

$$mv = mu + Ft$$

i.e.
$$Ft = mv - mu.$$

Momentum

A force F, acting for time t has produced a change Ft in this quantity mass × velocity. The momentum of a body is defined

$$\text{Momentum} = \text{mass} \times \text{velocity}$$

so that a force F acting for time t produces a change Ft in the momentum of a body.

Since momentum measures the effect of a force acting for a certain interval of time the units of momentum in SI units are newton seconds (N s). A force of 6 N acting for 5 s produces a change of 30 N s in the momentum of a body: a body mass 7 kg velocity 3 m s^{-1} has momentum of 21 N s.

The quantities F, u and v are vectors, of course, so that if F and u are not in the same direction the equation should be written

$$\mathbf{v} = \mathbf{u} + \frac{\mathbf{F}}{m} t,$$

and the momentum equation is

$$t\mathbf{F} = m\mathbf{v} - m\mathbf{u}.$$

The convention for multiplication of a vector by a scalar is to write the scalar in front of the vector, so the order of the terms F and t has been changed. Thus a force $(4\mathbf{i} + 6\mathbf{j})$ N acting for 5 s produces a change of $(20\mathbf{i} + 30\mathbf{j})$ N s in the momentum of a body: a body mass 7 kg velocity $(2\mathbf{i} + 3\mathbf{j})$ m s^{-1} has momentum $(14\mathbf{i} + 21\mathbf{j})$ N s.

Impulse

The effect of the force $\mathbf{F}$ acting for time t is to produce a change $t\mathbf{F}$ in the momentum of the body. This is called the impulse, the definition being

Impulse = force × time.

It often happens that we consider a large force acting for a very small interval of time, as when a cricket ball strikes a bat, and we may speak of the ball receiving an impulse, implying that the impulse is given in a very small period of time. There is, of course, no need for an impulse to be restricted to a short interval of time. A force 10 N acting for 5 s has an impulse of 50 N s: a force 10^4 N acting for 5×10^{-3} s also gives an impulse of 50 N s.

A body mass 4 kg, initial velocity $(3\mathbf{i} + 2\mathbf{j})$ m s^{-1}, later velocity $(5\mathbf{i} - 6\mathbf{j})$ m s^{-1}, has received an impulse $4(5\mathbf{i} - 6\mathbf{j}) - 4(3\mathbf{i} + 2\mathbf{j})$ N s, i.e. an impulse of $(8\mathbf{i} - 32\mathbf{j})$ N s, whatever the interval of time which elapsed between the measurement of these velocities.

From the momentum of a body we can find the change in the velocity caused by a known force, or we can find the force necessary to cause a certain change in velocity.

Example 1. *A body mass 2 kg initially at rest has velocity 36 m s^{-1} after a constant force has acted on it for 3 s. Find the force.*

The initial momentum of the body was 2×0, i.e. 0.

The final momentum is 2×36, i.e. 72 N s.

Hence the change in momentum is 72 N s. This was caused by a force X newtons acting for 3 s; thus

$$3X = 72$$
$$\Rightarrow X = 24,$$

i.e. the force is 24 N.

Example 2. *A body mass 5 kg with velocity 2 m s^{-1} is acted on by a force of 15 N for 4 s. Find the final velocity of the body.*

The initial momentum of the body is 10 N s.

The change in momentum of the body is 4×15, i.e. 60 N s.

Hence the final momentum is 70 N s.

If the final velocity is v m s^{-1},

$$5v = 70$$
$$\Rightarrow v = 14.$$

The final velocity is 14 m s^{-1}.

This working can be presented more precisely

$$4 \times 15 = 5v - 5 \times 2$$
$$\Rightarrow 5v = 70$$
$$\Rightarrow v = 14.$$

It is important to think of momentum as the physical quantity changed by a force in a certain time, as illustrated in the equations

$$Ft = mv - mu$$

and

$$t\mathbf{F} = m\mathbf{v} - m\mathbf{u}.$$

Example 3. *A body mass 20 kg velocity $(3\mathbf{i} + 4\mathbf{j})$ m s^{-1} is acted on by a force $(100\mathbf{i})$ N for 6 s. Find the subsequent velocity of the body.*

Using the equation $t\mathbf{F} = m\mathbf{v} - m\mathbf{u}$, we have

$$6(100\mathbf{i}) = 20(\mathbf{v}) - 20(3\mathbf{i} + 4\mathbf{j}),$$
$$\Rightarrow 20\mathbf{v} = 660\mathbf{i} + 80\mathbf{j}$$
$$\Rightarrow \mathbf{v} = 33\mathbf{i} + 4\mathbf{j}.$$

The subsequent velocity is $(33\mathbf{i} + 4\mathbf{j})$ m s^{-1}.

Notice that in this example there was not a component of force in the **j** direction; the momentum of the body in that direction was unaltered, therefore the velocity of the body in that direction was unaltered.

EXERCISE 18

1. Write down the momentum of each of the following:
   (a) a body mass 5 kg, velocity 20 m s^{-1},
   (b) an athlete mass 60 kg, velocity 6 m s^{-1},
   (c) a car mass 1000 kg, velocity 77 m s^{-1},
   (d) a cricket ball mass 0.14 kg, velocity 30 m s^{-1},
   (e) an electron mass 10^{-30} kg, velocity 10^8 m s^{-1}.
2. Write down the momentum of a body mass 4.8 kg, velocity $(2\mathbf{i} - 5\mathbf{j})$ m s^{-1}.
3. A body mass 10 kg initially at rest has velocity 5 m s^{-1} after a force F has acted on it for 4 s. Find F.
4. A body mass 10 kg initially at rest has velocity $(5\mathbf{i} + 2\mathbf{j})$ m s^{-1} after a force **F** has acted on it for 4 s. Find **F**.
5. A body mass 3 kg initial velocity $(2\mathbf{i} - \mathbf{j})$ m s^{-1} has velocity $(3\mathbf{i} + 5\mathbf{j})$ m s^{-1} after a force **F** has been acting on it for 6 s. Find **F**.
6. A body mass 10 kg initial velocity $(\mathbf{i} + 4\mathbf{j})$ m s^{-1} has velocity $(5\mathbf{i} + 2\mathbf{j})$ m s^{-1} after a force **F** has acted on it for 5 s. Find **F**.
7. A body mass 5 kg initially at rest is acted on by a force $(4\mathbf{i} + \mathbf{j})$ N for 3 s. Find the velocity of the body at the end of the 3 s.
8. A bird mass 0.4 kg initial velocity $(10\mathbf{i} + 12\mathbf{j})$ m s^{-1} has velocity $(7\mathbf{i} + 30\mathbf{j})$ m s^{-1} after an interval of 10 s. Find the force which has been acting on the bird.
9. A fives ball, mass 0.05 kg travelling with velocity 12 m s^{-1} strikes a wall at right angles and rebounds with velocity 10 m s^{-1}. Find the impulse given by the wall to the ball.
10. A squash ball mass 0.03 kg travelling with velocity $(10\mathbf{i} + 8\mathbf{j})$ m s^{-1} strikes a wall and rebounds with velocity $(-4\mathbf{i} + 8\mathbf{j})$ m s^{-1}. Find the impulse given by the ball to the wall and the impulse given by the wall to the ball.
11. A stone mass 0.2 kg is dropped over a cliff. Write down the force in newtons which acts on it, and hence find its momentum and velocity after 3 s.
12. A stone mass 0.05 kg is thrown over a cliff with velocity $(10\mathbf{i} + 5\mathbf{j})$ m s^{-1}, where **i** is a unit horizontal and **j** a unit vector vertically upwards. Find its velocity, in vector form, 3 s later.

13. A cricket ball mass 0.14 kg receives an impulse of 4 N s. Find the velocity it acquires.

14. A cricket ball mass 0.14 kg receives an impulse of $(2.1\mathbf{i} + 2.8\mathbf{j})$ N s. Find, in vector form, the velocity it acquires, and find the magnitude of this velocity.

15. A cricket ball mass 0.15 kg initial velocity $32\mathbf{i}$ m s^{-1} receives an impulse of $3.0\mathbf{j}$ N s. Find its subsequent velocity in vector form and the magnitude of the velocity.

16. A coin mass 0.005 kg is given a velocity of 4 m s^{-1} across a rough horizontal table. It comes to rest after 1 s. Find the force which has been acting on the coin and the coefficient of friction between the coin and the table.

Further applications of momentum

Variable forces

The ideas which we have introduced, the momentum of a body and the impulse which a force exerts over an interval of time can be extended further. First, we have assumed that the force is constant. This is not a necessary restriction but was merely an initial simplification. From Newton's Law

$$\mathbf{F} = m\mathbf{a}$$

and the calculus definition of acceleration 'the rate of change of velocity', we have

$$\mathbf{F} = m\frac{\mathrm{d}\mathbf{v}}{\mathrm{d}t}$$

Integrating over an interval of time, say t goes from t_1 to t_2,

$$\int_{t_1}^{t_2} \mathbf{F}\,\mathrm{d}t = m\mathbf{v} - m\mathbf{u}$$

$\mathbf{u}$ and $\mathbf{v}$ being the initial and final velocities as before. This extension to apply to variable forces is also useful when the expression for the force in terms of t is non-integrable, or not known, and approximate methods have to be used.

Example 1. *A body mass 5 kg initially at rest is acted on by a force $(5 + 2t)$ N, where t seconds is the time which has elapsed since the beginning of the motion. Find the velocity of the body after 4 s.*

Applying the above method,

$$\int_0^4 (5 + 2t)\mathrm{d}t = \int 5\frac{\mathrm{d}v}{\mathrm{d}t}\,\mathrm{d}t$$

$$\Rightarrow \left[(5t + t^2)\right]_0^4 = \left[5v\right]_0^V$$

where V m s^{-1} is the velocity after 4 s. Hence

$$36 = 5V$$

$$\Rightarrow V = 7.2.$$

The velocity of the body after the force has acted on it for 4 s is 7.2 m s^{-1}.

Example 1. *A body mass 2 kg initially at rest is acted on by a force whose value is given by the table below. Estimate the velocity of the body at the end of 6 s.*

t (seconds)	0	1	2	3	4	5	6
F (newtons)	0	0.8	1.6	2.0	2.5	2.6	2.7

Using the momentum equation

$$\int F\,\mathrm{d}t = mv - mu,$$

we have

$$\int_0^6 F\,\mathrm{d}t = 2V, \qquad\qquad (i)$$

where V m s^{-1} is the velocity after 6 s. We wish to find an approximate method of estimating the value of the integral.

From the calculus we know that the area bounded by the curve, the x-axis and two ordinates is equal to the value of the integral, and Fig. 115 shows that this area can be divided into trapeziums and summed in that manner.* It is important to emphasize that the force is probably NOT given by the straight lines joining the given points, but that is merely a convenient approximation. Thus

$$\begin{aligned}
\text{area} &= \tfrac{1}{2} \times 1 \times (0 + 0.8) + \tfrac{1}{2} \times 1 \times (0.8 + 1.6) \\
&\quad + \tfrac{1}{2} \times 1 \times (1.6 + 2.0) + \tfrac{1}{2} \times 1 \times (2.0 + 2.5) \\
&\quad + \tfrac{1}{2} \times 1 \times (2.5 + 2.6) + \tfrac{1}{2} \times 1 \times (2.6 + 2.7) \\
&= \tfrac{1}{2}(21.7) = 10.85.
\end{aligned}$$

*For a fuller discussion of this method, and also of Simpson's rule, see *Additional Pure Mathematics*, by L. Harwood Clarke.

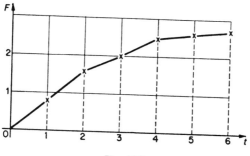

Fig. 115

Returning to (i) we see that

$$10.85 = 2V,$$

$$\Rightarrow V = 5.425.$$

The estimated final velocity is about 5.4 m s⁻¹.

Checking, the average force during the 6 s interval is about 2 N, so that the change in momentum will be about 12 N s. Our estimate for the velocity of about 5.4 m s⁻¹ gives a final momentum of 10.8 N s, which is reasonable.

Collisions

If two bodies, masses m_1 and m_2 moving in the same straight line, with velocities v_1 and v_2 respectively, collide, the force exerted by m_1 on m_2 will be equal and opposite to the force exerted by m_2 on m_1 (Newton's Third Law). These forces will act for the same interval of time so the change in the momentum of m_1 is equal and opposite to the change in the momentum of m_2. Therefore the total momentum of both bodies is unaltered by the collision.

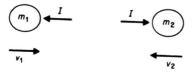

Fig. 116

In investigating all problems involving collisions, it is most important to draw a diagram marking clearly the directions of velocities and impulses, if the latter are required.

Momentum is conserved even if the bodies do not move in the same straight line. These problems are beyond the scope of this book, but are treated in substantially the same manner.

Example 1. *A body of mass 3 kg velocity 4 m s⁻¹ overtakes and collides with a body mass 2 kg moving in the same direction with velocity 3 m s⁻¹. If the bodies move away together, find their velocity.*

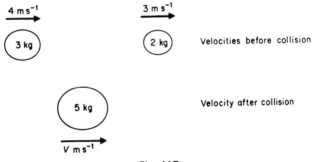

Fig. 117

The total momentum before the collision is

$$3 \times 4 + 2 \times 3 = 18 \text{ N s}$$

The momentum after the collision is $5V$ N s. Therefore

$$V = 3.6.$$

The bodies have a common velocity of 3.6 m s⁻¹ after the collision.

In most problems involving impacts or collisions, we think of there being two unknowns, the velocities of each body after the collision. Later mechanics will show how to solve such problems, but for the present we have to restrict ourselves to problems with some simplification, such as the bodies moving off with the same velocity, or when we have an extra piece of data about the final velocities.

Example 2. *A body mass 1 kg moving with velocity (5i) m s⁻¹ collides with a body mass 2 kg velocity −4i m s⁻¹. The second body is brought to rest by the collision. Find the velocity of the first body after the collision.*

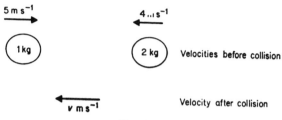

Fig. 118

The total momentum before the collision is

$$1 \times 5 + 2 \times (-4) = -3 \text{ N s.}$$

The total momentum after the collision is

$$1 \times (-v) + 2 \times 0 = -v \text{ N s}$$
$$\Rightarrow -v = -3$$

i.e.
$$v = 3.$$

The velocity after collision of the first body is $-3i$ m s⁻¹.

Forces on surfaces due to loss of momentum

We have probably all lain awake at night and heard the tremendous force with which wind can rattle windows, and even wondered whether the force of a gale could break the window-pane. Even more tangible is the force with which a jet of water from a hose can strike us! In both these examples momentum is being destroyed in any given interval of time, so a force must be exerted on the air or on the stream of water. Hence an equal and opposite force will be exerted by the air on the window-pane, and by the jet of water on us.

Think of the force on a window, say 1.5 m by 1.2 m, when wind travelling at 20 m s⁻¹ is striking it. Let us assume that the wind is brought to rest by the window. Every second the volume of air which strikes the window is

$$(1.5 \times 1.2) \times 20 \text{ m}^3 = 36 \text{ m}^3.$$

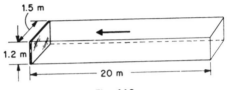

Fig. 119

The density of air may be taken as 1.3 kilogrammes per cubic metre, so the mass of air brought to rest each second is

$$36 \times 1.3 = 46.8 \text{ kg}.$$

The velocity of the air is 20 m s⁻¹, so the momentum destroyed each second is

$$46.8 \times 20 = 936 \text{ N s}.$$

If F is the force in newtons exerted by the window on the air, considering the change in momentum each second,

$$F.1 = 46.8 \times 20 - 46.8 \times 0$$
$$F = 936.$$

The force on the window is 936 N. This force is spread over an area 1.5 m × 1.2 m, so the force per m² is

$$\frac{936}{1.8} \text{ N} = 520 \text{ N}.$$

A quick repetition of the calculation will show that if the speed of the wind increases by a factor of 2 to 40 m s⁻¹, the force per square metre is increased by a factor 4 to 2080 N. The force per unit area is proportional to the square of the velocity of the wind. It is hardly surprising that storms can do so much damage.

EXERCISE 19

1. A body mass 3 kg velocity 5 m s^{-1} strikes a body mass 2 kg initially at rest. If the bodies move away together, find their common velocity.

2. A body mass 2 kg velocity 4 m s^{-1} strikes a body mass 3 kg moving with velocity 4 m s^{-1} in the opposite direction. If the heavier body is brought to rest by the collision, find the velocity of the lighter body after the collision.

3. A ball mass 0.5 kg velocity 10 m s^{-1} strikes an identical ball initially at rest. If they separate with one ball moving 5 m s^{-1} faster than the other, but both travelling in the same direction in the same straight line, find the velocity of each ball after the impact.

4. A body mass 3 kg travelling with velocity 5$\mathbf{i}$ m s^{-1} catches up a body mass 2 kg travelling with velocity 2.5$\mathbf{i}$ m s^{-1}. If they coalesce, find their subsequent velocity.

5. A body mass 4 kg travelling with velocity 5$\mathbf{i}$ m s^{-1} impinges directly on a body of mass 6 kg travelling with velocity $-3\mathbf{i}$ m s^{-1}. If they coalesce, find their subsequent velocity.

(More questions on impacts will be found in Exercise 20.)

6. A jet of water issues from the nozzle of a hose, area of cross-section 10^{-4} m^2, with velocity 10 m s^{-1}. Find the force per square metre if all the water strikes a person, surface area struck by water 0.2 m^2. Assume that the velocity of the water is not reduced before the impact, and that all the momentum of the water is destroyed on impact.

7. Sea water, density 1.03 × 10^3 kg m^{-3}, strikes a sea wall with velocity 20 m s^{-1}. Calculate the force per square metre on the wall.

8. A gardener holds a length of hose so that it is in the shape of a quadrant of a circle. The area of the cross-section of the hose is 4 × 10^{-4} m^2. Water flows through the hose at a steady speed of 10 m s^{-1} and issues at right angles to the path along which it entered the hose. By considering the change of momentum of the water, find the force which the gardener exerts on the hose.

9. The force F newtons acting on a body mass 5 kg is given by

$$F = 6 - 2t,$$

where t seconds is the time the force has been acting. If the body is at rest when $t = 0$, find the velocity of the body (a) after 2 s, (b) after 5 s.

When is the maximum velocity attained? Find this maximum velocity.

10. The force F newtons acting on a body mass m kg is given by

$$F = a - bt,$$

where t seconds is the time the force has been acting. If the body is at rest when $t = 0$, find the velocity of the body when $t = T$.

11. The force F newtons acting on a body mass 4 kg at time t seconds from the start is given by the following table:

t	0	1	2	3	4	5	6
F	0	1.1	2.4	4.1	6.2	8.0	9.0

Estimate the velocity of the body when $t = 6$.

12. The force F newtons acting on a body mass 5 kg at time t seconds from the start is given by the following table:

t	0	1	2	3	4	5	6	7	8
F	0	0	1.3	3.4	4.6	4.7	5	5	5

Estimate the velocity of the body when (a) $t = 6$, (b) $t = 8$.

CHAPTER 12

Energy and Work

Kinetic energy

In the previous chapter we saw that the effect of a force $\mathbf{F}$ acting on a body for time t was to produce a change $t\mathbf{F}$ in a quantity $m\mathbf{v}$ which we defined as the momentum of the body. In particular, when $\mathbf{F}$ is measured in newtons and t in seconds, as in SI units, the change in the momentum of the body is $t\mathbf{F}$ newton seconds (N s). Can we find the effect of a force $\mathbf{F}$ acting through a distance s, and is there any profit in doing so?

Consider first a constant force F, and assume that all motion is in the same straight line as the line of action of the force F. Then the acceleration which this force produces in a body mass m is F/m, and if this takes place over a distance s, the equation of motion linking the final velocity v with the initial velocity u is

$$v^2 = u^2 + 2as,$$

hence
$$v^2 = u^2 + 2\frac{F}{m}s.$$

Since we are interested in the effect of F over a distance s we can rewrite this

$$Fs = \tfrac{1}{2}mv^2 - \tfrac{1}{2}mu^2.$$

A change Fs has been produced in the quantity

$$\tfrac{1}{2}(\text{mass}) \times (\text{velocity})^2$$

and this quantity we call the *kinetic energy* of the body. The kinetic energy of a body mass m velocity v is defined as $\tfrac{1}{2}mv^2$. In SI units, kinetic energy is measured in joules, and since all SI units are consistent, a body mass 1 kg, velocity 1 m s^{-1} has kinetic energy $\tfrac{1}{2} \times 1 \times 1^2$ joules, or $\tfrac{1}{2}$ joule (abbreviated $\tfrac{1}{2}$ J).

Since F, s and v are vectors, we wonder whether this can be extended to motion in two directions, or to problems in which the force is not acting along the line of initial motion of the body. The product of two vectors is needed here, and this is just outside the scope of this present book, but the definition is still valid and all our results can be extended to problems in two or even three dimensions.

As with momentum, the force F need not be constant. The equation of motion

$$F = m \frac{\mathrm{d}v}{\mathrm{d}t}$$

is expressed in the form

$$F = m \frac{\mathrm{d}v}{\mathrm{d}s} \cdot \frac{\mathrm{d}s}{\mathrm{d}t}$$

i.e.
$$F = mv \frac{\mathrm{d}v}{\mathrm{d}s}.$$

Integrating,
$$\int F \, \mathrm{d}s = \left[\tfrac{1}{2}mv^2 \right]_u^v$$

$$\int F \, \mathrm{d}s = \tfrac{1}{2}mv^2 - \tfrac{1}{2}mu^2.$$

The following three examples illustrate very simple applications of kinetic energy.

Example 1. *A body mass 4 kg moving initially with velocity 6 m s^{-1} has its velocity reduced to 3 m s^{-1}. Find the loss of kinetic energy.*

$$\text{Initial kinetic energy} = \tfrac{1}{2} \times 4 \times (6)^2$$
$$= 72 \text{ J}.$$
$$\text{Final kinetic energy } = \tfrac{1}{2} \times 4 \times (3)^2$$
$$= 18 \text{ J}.$$
$$\therefore \qquad \text{Loss of kinetic energy} = 54 \text{ J}.$$

Example 2. *A body mass 4 kg moving initially with velocity 6 m s^{-1} is acted upon by a force of 20 N over a distance of 22 m. Find the final velocity of the body.*

Initial kinetic energy is 72 J as before. The effect of the force

20 N over 22 m is to increase the kinetic energy by 20 × 22. Hence the increase of kinetic energy is 440 J. From

$$Fs = \tfrac{1}{2}mv^2 - \tfrac{1}{2}mu^2,$$

we get for the final kinetic energy (72 + 440) J. Hence

$$\tfrac{1}{2}mv^2 = 512,$$
$$\Rightarrow \tfrac{1}{2}(4)v^2 = 512,$$
$$\Rightarrow v = 16,$$

i.e. the final velocity is 16 m s⁻¹.

Example 3. *A bullet mass 0.02 kg, initial velocity 100 m s⁻¹, has its velocity reduced to 40 m s⁻¹ when it has gone 0.4 m into sand. Find the resisting force exerted by the sand.*

Initial kinetic energy = $\tfrac{1}{2}$ × (0.02) × (100)² = 100 J.

Final kinetic energy = $\tfrac{1}{2}$ × (0.02) × (40)² = 16 J.

If the resisting force is F newtons,

$$F \times (0.4) = 100 - 16,$$
$$\Rightarrow F = 210,$$

the resisting force of the sand is 210 N.

Work

The force 210 N in the last example did some work on the bullet; it reduced the kinetic energy of the bullet from 100 J to 16 J. The bullet travelled 0.4 m while this energy was being reduced. A greater force, say 2100 N, would have reduced the energy in a shorter distance, 0.04 m: a smaller force, say 21 N, would have needed a greater distance, 4 m. The amount by which the kinetic energy has been reduced is the work done by the force, and *work* is defined as the product *force × distance, in the direction of the force*. Here again, when we discover later how to find the product of two vectors, a vector definition is possible and desirable.

Work is not only done in increasing the kinetic energy of a body. If we stretch an elastic spring we shall do work, yet there will not be an increase in the velocity of the spring, so there will not be an increase in the kinetic energy of the spring. The energy in this case is stored as elastic energy in the spring. Similarly, a bricklayer raising a bucketful of bricks at a steady speed will be doing work, yet there will be no increase in the speed of the bucketful of bricks. The work here has been stored as (gravitational) *potential energy*, and in this case could easily be converted into kinetic energy by letting go the rope! Subject to a certain loss of energy because of the inefficiency of the pulley, if the rope passes over a pulley the potential energy would be rapidly converted into kinetic energy, to measure the speed with which the bucket struck the ground. Many other forms of energy are studied in physics, such as magnetic and electrostatic potential energy, heat, sound and chemical energy.

Rate of working: power

A machine projecting cricket balls, mass 0.14 kg each, with velocity 20 m s^{-1} gives to each cricket ball kinetic energy

$$\tfrac{1}{2}(0.14) \times 20^2 \text{ J} = 28 \text{ J}.$$

If it projects one such cricket ball every second, it is doing 28 joules of work per second. If it throws five cricket balls each second, it is doing 140 joules per second. The rate at which the machine works is called the *power* of the machine. If the machine in this example increases its rate of working to project ten cricket balls each second, then the power needed is 280 joules per second.

The unit for measuring power in SI is the *watt* (W), defined as 1 joule per second. The machine was first working at 28 W, then 140 W, finally 280 W. Because the watt is an inconveniently small unit, the kilowatt is more commonly used, where

$$1 \text{ kilowatt (kW)} = 1000 \text{ watts.}$$

The name watt is of great historical interest. James Watt (1736–1819) was a very famous engineer responsible amongst other things for harnessing the power of steam. His unit of work was called the horse-power (1 horse-power $\simeq$ 0.746 kW) and was based on observations of the rate at which the large horses once used by brewers worked. It is about 10 per cent greater than the rate at which most horses can work. The term cheval-vapeur is still used in many European countries, and 1 cheval-vapeur is slightly less than 1 horse-power.

It must be emphasized that the work done by a force depends only on the component of the force in the direction in which motion takes place. (This is later brought out clearly by the vector definition of the work done.) If a body mass 10 kg is to be raised 5 m vertically at a steady speed, a force 10 × 9.8 N (equal to the weight of the body) must act on it over a distance of 5 m. Thus the work done is 98 × 5 joules, i.e. 490 J.

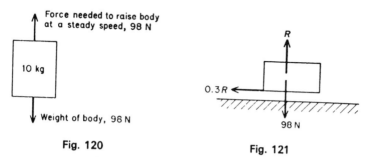

Fig. 120

Fig. 121

If the same body is to be dragged over a *smooth* horizontal floor at a constant velocity, then the force resisting motion is zero, so that a force is not necessary to maintain steady motion, and so work is not done. In practice the floor would be rough, however small the coefficient of friction might be in some cases. Let us suppose that the coefficient of friction is 0.3 in this example.

Considering the forces shown on the body in Fig. 121, resolving vertically, the normal reaction $R = 98$ N. Hence the frictional force is 29.4 N and the work done in moving the body 5 m over this rough floor is 29.4 × 5 joules, i.e. 147 J.

If the body being raised vertically had a velocity of 5 m s^{-1}, the work done in each second would be 490 J, so that the power necessary to raise it at this speed would be 490 W, i.e. 0.49 kW. If the velocity was 20 m s^{-1}, the body would rise 20 m each second, so that the work done would be 98 × 20 J and the power needed would be 1.96 kW. The body moving horizontally over that rough floor, by contrast, would only need 0.147 kW to travel at 5 m s^{-1} and 0.588 kW to travel at 20 m s^{-1}.

In problems in which the body is not moving in the same straight line as the line of action of the forces acting on it, it is almost invariably desirable to consider the components of the forces in the direction in which motion takes place, as the following example shows.

Example. *A body mass 10 kg has velocity 20 m s⁻¹ at the foot of a rough plane inclined at 30° to the horizontal, and 2 m s⁻¹ when it has travelled 20 m up the plane. Find the average frictional force exerted by the plane on the body.*

There are three forces acting on the body; the frictional force which is in the line of motion, the normal reaction which is at right angles to the line of motion and so against which no work is done, and the weight of the body which is at 60° to the line of motion of the body. Replace the weight of the body by the components of the weight along and perpendicular to the line of motion, as in Fig. 122(a) and (b).

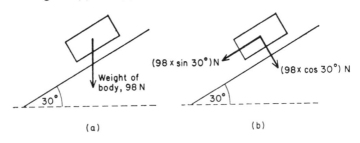

Fig. 122

The forces acting on the body are now shown in Fig. 123, and we can see clearly against which forces work will be done. The weight of the body has been replaced by the two components (98 cos 30°) N into the plane, which we do not need to evaluate, and (98 sin 30°) N along the plane, which equals 49 N.

The kinetic energy of the body initially is $\frac{1}{2}(10) \times (20)^2$ J $= 2000$ J.

The kinetic energy of the body finally is $\frac{1}{2}(10) \times (2)^2$ J $= 20$ J.

The total forces opposing motion $= (49 + F)$ newtons. Hence the work done in travelling 20 m along the plane is $(49 + F) \times 20$ J.

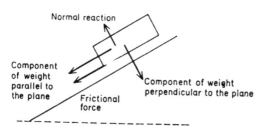

Fig. 123

Since the work done by the body is equal to the loss of kinetic energy,

$$(49 + F) \times 20 = 2000 - 20$$

i.e.
$$F = 50.$$

The frictional force exerted by the plane on the body is 50 N.

Loss of energy when bodies collide

When a bullet travelling with quite a high velocity, say 300 m s^{-1}, penetrates a building, all the kinetic energy of the bullet is lost as the bullet is brought to rest. When a bullet penetrates and becomes embedded in a body which is free to move, they will move away together. Has energy been lost in this case?

In the previous chapter we saw that momentum was conserved in collisions, because the force of the bullet on the block was equal and opposite to the force of the block on the bullet. As they act for the same interval of time, the momentum of both together is unaltered. But considering the work done by the bullet on the block and by the block on the bullet, although the forces are equal and opposite the distance moved by the bullet is not the same as the distance moved by the block, so that *energy is not generally conserved in impacts*.

Suppose that the bullet has a mass of 0.01 kg, velocity 300 m s^{-1} and that it becomes embedded in a block mass 50 kg.

From the conservation of momentum,

$$300 \times 0.01 = 50.01 \, v$$

where v m s^{-1} is the velocity of the bullet and block together. Hence,

$$v \simeq 0.06,$$

i.e. the velocity of the bullet and block together is about 0.06 m s^{-1}.

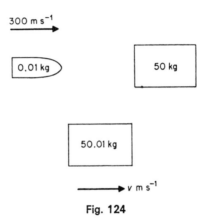

Fig. 124

Now the kinetic energy of the block before impact is 0, since the block is at rest, and the kinetic energy of the bullet before impact is

$$\tfrac{1}{2} \times (0.01) \times (300)^2 = 450 \text{ J}.$$

The kinetic energy of the bullet and block together is

$$\tfrac{1}{2} (50.01) \times (0.06)^2 \simeq 0{\cdot}09 \text{ J}.$$

We see that nearly all the kinetic energy of the bullet has been lost by the impact, only 0.09 J remaining out of 450 J before the impact. If the masses of the bodies had not been so very different in size, or if the bullet had not become embedded in the block, the loss of kinetic energy would not have been so great.

To find the power or rate of working of a force

Power is the rate at which work is done, and in investigating the power produced by a machine or the power required to do certain work it is important to consider how much work is being done each second. A common example is the one we discussed earlier, a mass of 10 kg being raised vertically or being dragged over rough horizontal ground.

A more difficult problem occurs when a machine overcomes a force and gives kinetic energy to a body, as when a pump raises water from a well and discharges it with a given velocity. The following example shows the method of procedure.

Example *A pump raises 0.1 m³ of water a second through 12 m and discharges it with velocity 8 m s⁻¹. Find the power required.*

The work done per second consists of two parts; the water is raised through 12 m, and is also given some kinetic energy.

The mass of water raised each second is 100 kg, since the mass of 1 m³ of water is 1000 kg. The weight of this water is 9.8×100 N. Hence the work done per second against the weight of the water is

$$9.8 \times 100 \times 12 \text{ J} = 11\ 760 \text{ J}.$$

The kinetic energy given each second to the water is

$$\tfrac{1}{2} \times 100 \times (8)^2 \text{ J} = 3200 \text{ J}.$$

Hence the total work done per second is

$$(11\ 760 + 3200) \text{ J} = 14\ 960 \text{ J}.$$

The rate at which the pump is working is 14.96 kW.

Tractive force

A common situation in which power is used is to drive a car. If the engine of a car is working at 30 kW, the car could travel at 30 m s⁻¹ against resistances of 1000 N, at 15 m s⁻¹ against resistances of 2000 N, at 10 m s⁻¹ against resistances of 3000 N, and so on. What happens if the resistances total 1000 N, the car is travelling at 10 m s⁻¹ and the engine is working at 30 kW, greater than needed merely to overcome the resistance?

It is convenient here to introduce the term *tractive force*, and this may be thought of as equal in magnitude to the greatest resistance the car could overcome at a certain speed. Notice that the tractive force decreases as the speed increases, if the power remains constant. For this car, the tractive force is 1000 N at 30 m s^{-1}, 2000 N at 15 m s^{-1} and 3000 N at 10 m s^{-1}.

An engine power P watts in a car velocity v m s^{-1} produces a tractive force of P/v newtons.

When the tractive force is 3000 N, as in this case, and the resistances are only 1000 N, there is a force 'to spare' of 2000 N. This produces acceleration. If the mass of this car is 1000 kg,

at 30 m s^{-1}, 'spare force' = 0 $\Rightarrow$ acceleration = 0,

at 15 m s^{-1}, 'spare force' = 1000 N $\Rightarrow$ acceleration = 1 m s^{-2},

at 10 m s^{-1}, 'spare force' = 2000 N $\Rightarrow$ acceleration = 2 m s^{-2}.

If the resistances are larger than the tractive force, the car of course slows down.

Example 1. *A car mass 750 kg has an engine which produces a constant 40 kW. If the resistance to motion, assumed constant, is 800 N, find the maximum speed at which the car can travel on level road, and the acceleration when it is travelling at 25 m s^{-1}.*

The tractive force is $40\,000/v$ newtons. At maximum speed, this is equal to the resistances, i.e. $40\,000/v = 800$

$$v = 50.$$

The maximum speed is 50 m s^{-1}.

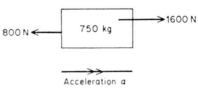

Fig. 125

At 25 m s^{-1}, the tractive force is $40\,000/25 = 1600$ N. From $F = ma$ we get

$$1600 - 800 = 750a$$

$$\Rightarrow a = \tfrac{800}{750} = \tfrac{16}{15}.$$

The acceleration at 25 m s^{-1} is $\tfrac{16}{15}$ m s^{-2}.

Example 2. *A car mass 1000 kg is travelling up a hill of 1 in 49, with the engine working at a constant 40 kW. Find the greatest speed the car can attain up that hill, and also the acceleration when the car is travelling up the hill at 25 m s⁻¹, if there are resistances to motion of 800 N.*

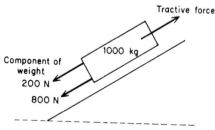

Fig. 126

As the car is going up a hill of 1 in 49, there is a component

$$\tfrac{1}{49} \times 1000 \times 9.8 = 200 \text{ N}$$

of the weight of the car acting down the hill.

$$\text{Tractive force} = \frac{40\,000}{v} \text{ newtons.}$$

At maximum speed,

$$\frac{40\,000}{v} = 800 + 200$$

$$\Rightarrow v = 40.$$

The maximum speed up the hill is 40 m s⁻¹.

When the car is travelling at 25 m s⁻¹, the tractive force is

$$\frac{40\,000}{25} \text{ N} = 1600 \text{ N.}$$

From $F = ma$ we get

$$1600 - (800 + 200) = 1000a$$

$$\Rightarrow a = \frac{600}{1000} = 0.6.$$

The acceleration at 25 m s⁻¹ is 0.6 m s⁻².

In practice the power, and the air and other resistances are rarely constant, and usually vary with the speed of the car. This requires the use of calculus in dealing with these problems.

Units. The unit of energy is the joule (J): the unit of power is the watt (W).

The kinetic energy of a body mass m kg, velocity v m s^{-1} is $\frac{1}{2}mv^2$ joules.

The work done by a force of F newtons over a distance of s metres in the direction of the force is Fs joules.

The power which provides 1 joule of energy per second is 1 watt (W).

EXERCISE 20

1. Find the kinetic energy of each of the following:
   (a) a boy of mass 50 kg running at 6 m s^{-1},
   (b) a car of mass 500 kg travelling at 30 m s^{-1},
   (c) a hovercraft of mass 10^5 kg travelling at 25 m s^{-1},
   (d) an insect of mass 1 g, velocity 1 m s^{-1},
   (e) an electron, mass 9×10^{-28} g, velocity 2×10^8 m s^{-1}.
2. A body of mass 25 kg has kinetic energy 800 J. What is the velocity of the body?
3. A bullet moving with velocity 400 m s^{-1} has kinetic energy 800 J. What is the mass of the bullet?
4. Find the kinetic energy of a body initially at rest after a force of 80 N has pushed it 0.6 m.
5. If the body in question 4 has a mass of 12 kg, what is its final velocity?
6. Find the increase in energy when a force of 5.2 N acts on a body over a distance of 4.8 m.
7. Find the work done in raising a body of mass 50 kg a distance of 80 m.
8. Find the work done in raising a body of mass 50 kg a distance of 8 m into a space craft stationary on the surface of the moon. (The acceleration due to gravity on the moon's surface may be taken to be 1.65 m s^{-2}.)
9. Find the work done in raising a load of bricks, mass 5000 kg, a distance of 24 m.
10. Find the work done in raising 1 m^3 of water a distance of 25 m.
11. Find the work done in raising a body of mass 20 kg a distance of 40 m.

12. Find the work done in dragging a body of mass 20 kg a distance of 40 m over a smooth horizontal floor.

13. Find the work done in dragging a body of mass 20 kg a distance of 40 m over a rough horizontal floor, the coefficient of friction between the body and the floor being 0.6.

14. Find the work done in dragging a body of mass 20 kg a distance of 40 m over a smooth plane inclined at 30° to the horizontal.

15. Find the work done in dragging a body of mass 20 kg a distance of 40 m up a rough plane inclined at 30° to the horizontal (the coefficient of friction is 0.5).

16. Find the work done in giving a body of mass 25 kg a velocity of 8 m s⁻¹.

17. Find the work done in giving 1 m³ of water a velocity of 8 m s⁻¹.

18. Find the work done in giving 1 m³ of petrol a velocity of 16 m s⁻¹. (The mass of 1 m³ of petrol is 800 kg.)

19. A man of mass 80 kg falls off the top of a building 50 m high. Neglecting air resistance, what is the increase in his kinetic energy? What is his final velocity?

20. A body of mass 10 kg increases velocity from 10 m s⁻¹ to 20 m s⁻¹ while covering a distance of 5 m. What force, assumed constant, has been acting on it?

21. A car of mass 700 kg accelerates from rest to 20 m s⁻¹ while going 80 m. What force has produced this change in velocity?

22. A toboggan of mass 20 kg is pushed with a force of 100 N over a distance of 25 m. If the resistance to motion is 20 N, what increase in kinetic energy is produced? If the velocity was initially 5 m s⁻¹, what is the final velocity?

23. A boy of mass 40 kg reaches the horizontal part of a slide travelling at 3 m s⁻¹. If the resistance to motion is 120 N, how far will he travel before coming to rest?

24. A girl of mass 40 kg slides down a water chute 4 m long. If the total accelerating force is 200 N, with what velocity will she leave the chute if she started from rest?

25. A fives ball of mass 0.08 kg is dropped from a height of 1.6 m and rebounds to a height of 0.6 m. How much energy was lost by the impact?

26. The mass of a cyclist and his machine is 100 kg. When he is not pedalling he finds that his speed drops from 5 m s⁻¹ to 4 m s⁻¹ while travelling 10 m. Find the average resistance.

27. The same cyclist now pedals for 10 m and increases his speed from 4 m s⁻¹ to 5 m s⁻¹ while travelling 10 m. Assuming the resistances to motion are unaltered, find the average force he exerts.

28. When the cyclist in questions 26 and 27 raises his speed from 5 m s^{-1} to 6 m s^{-1} while travelling 10 m, what is the average force he exerts?

29. A book of mass 1 kg is dropped from a window 10 m above the ground. What is its kinetic energy when it strikes the ground? What is then the velocity of the book?

30. A pillow of mass 10 kg is dropped from a window 10 m above the ground. The air exerts a resistance of 13.5 N. What is the kinetic energy of the pillow when it strikes the ground? What is then the velocity of the pillow?

31. A body of mass 4 kg is dropped into liquid. The velocity at the surface of the liquid is 10 m s^{-1}, and when the body has sunk 4 m the velocity is 12 m s^{-1}. Find the resisting force which the liquid exerts on the body.

32. A parachutist of mass 80 kg is falling at 8 m s^{-1} when 1000 m above the ground, and 9 m s^{-1} when he reaches the ground. What is the resisting force exerted on him by the air? (Take $g = 9.807$ m s^{-2}.)

33. A boy of mass 40 kg slides down a chute inclined at 60° to the horizontal. If the chute is smooth and the boy starts from rest, with what velocity does he pass a point on the chute 10 m from his starting point?

34. If the chute in question 33 had been rough and exerted a constant frictional force of 60 N, with what velocity would he have passed the point 10 m from his starting-point?

35. A cyclist, mass 90 kg, together with his machine has a velocity of 10 m s^{-1} at the foot of a hill. He cycles 1000 m and rises 150 m. If his final velocity is 6 m s^{-1} what is the average force he has exerted?

36. A bullet of mass 0.01 kg velocity 600 m s^{-1} is brought to rest by protective sandbags. Find the kinetic energy lost. If the bullet entered 2.4 m into the sand, what is the average resisting force of the sandbags?

37. A body of mass 2 kg velocity 4 m s^{-1} strikes an identical body at rest. They move away together. Find the kinetic energy lost by the impact. What happens to this energy?

38. A body of mass 2 kg velocity 4 m s^{-1} strikes a body mass 3 kg moving in the same direction with velocity 3 m s^{-1}. They move away together. Find the kinetic energy lost by this impact.

39. A body of mass 2 kg velocity 4**i** m s^{-1} strikes a body mass 3 kg velocity -3**i** m s^{-1}. They move away together. Find the kinetic energy lost by the impact.

40. A body of mass 2 kg velocity 4**i** m s^{-1} strikes a body mass 3 kg velocity -7**i** m s^{-1}. After the impact the velocity of the 2 kg body is $\frac{1}{2}$ m s^{-1} less than the velocity of the 3 kg body. Find the kinetic energy lost by the impact.

41. Two bodies of mass 2 kg and 1 kg are moving in the same straight line. The heavier body has a velocity of 15 m s^{-1}. It catches up and adheres to the other body. They move on together with a velocity of 12 m s^{-1}. Find the initial velocity of the lighter body and the kinetic energy lost in the impact.

42. A crane raises a 5000 kg steel girder at 0.4 m s^{-1}. At what rate is the engine of the crane working? (Assume work is not lost in driving the crane.)

43. Digging a trench in his garden a man lifts 10 m^3 of soil through 1.4 m in one hour. If the density of the soil relative to water is 1.1, find the rate at which the man is working.

44. Find the power necessary for a train of mass 4×10^5 kg to travel at 80 km h^{-1} against resistances of 60 N per 10^3 kg.

45. A blacksmith wields a hammer of mass 30 kg, delivering 5 blows a minute on an iron bar. The speed of the hammer just before each blow is 10 m s^{-1} and the hammer is brought to rest by each blow. Find the rate at which the blacksmith is working.

46. A train of mass 5×10^5 kg is travelling at 30 m s^{-1} up a slope of 1 in 100. The frictional resistance is 50 N per 10^3 kg. Find the rate at which the engine is working.

47. A machine for firing clay pigeons throws 3 'birds' a minute. The mass of each 'bird' is 0.08 kg and the velocity with which each leaves the machine is 20 m s^{-1}. Find the power necessary to drive the machine assuming half of it is lost in the machine.

48. Find the kW used by a firepump which raises water a distance of 4 m and delivers 0.12 m^3 a minute at a speed of 10 m s^{-1}.

49. A car mass 800 kg ascends a hill of 1 in 10 at 20 m s^{-1}, the air resistance being 200 N. What power is the engine producing?

50. A light motorcycle whose mass including rider is 200 kg can go at 10 m s^{-1} up a plane of inclination 1 in 14 and at 20 m s^{-1} down the same plane. If the resistance varies as the square of the speed and the power developed by the machine is constant, find the power developed.

CHAPTER 13

Addition of Velocities: Relative Velocity

Addition of velocities

If we are able to row a boat at 4 km h^{-1}, and we try to row up-stream against a current of 1 km h^{-1}, then our rate of progress measured along the banks of the river will be 3 km h^{-1}. If we turn around and row downstream, we shall progress at 5 km h^{-1}. What happens if we aim the boat across the stream?

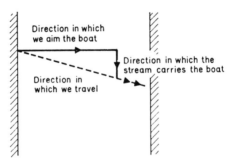

Fig. 127

If we were to row for one hour in this direction we should go 4 km at right-angles to the banks. The stream, however, would carry us 1 km downstream, so the path along which we should actually travel would be the path shown as — — → → in Fig. 127. Notice that at the end of the hour we should have covered more than 4 km but less than (4 + 1) or 5 km. By Pythagoras' theorem, from Fig. 128 we see we should be $\sqrt{17}$ km from our starting-point, about 4.1 km, so our velocity would be 4.1 km h^{-1}, correct to 2 significant figures.

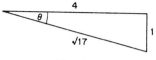

Fig. 128

The path along which we travel is not directly across tne river, but makes an angle θ with the line perpendicular to the banks of the river. By trigonometry,

$$\tan \theta = \tfrac{1}{4}$$
$$\Rightarrow \theta = 14°,$$

to the nearest degree.

The velocity we have and the direction of our path could, of course, be found by accurate drawing. We shall see that all these problems can be solved by drawing or by calculation. As a general rule, very simple problems with an obvious right-angled triangle are to be solved by calculation, but all others are more easily solved by drawing.

Resultant velocity

Our total velocity in this problem was the sum of the velocity which we gave to the boat and the velocity which the stream gave to the boat. Since all velocities are vectors, they were added as vectors.

If we set up coordinate axes as in Fig. 129 with Ox across the river and Oy along one bank, then with our usual notation,

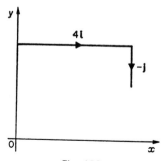

Fig. 129

the velocity with which we row $= 4\mathbf{i}$ km h^{-1},
the velocity of the stream $= -\mathbf{j}$ km h^{-1},
$\Rightarrow$ our total velocity $= (4\mathbf{i} - \mathbf{j})$ km h^{-1}.

The magnitude and direction of this are obtained immediately.

The advantages of vector notation become apparent as the problems become more complicated, but even problems not expressed in terms of unit vectors can be tackled quite easily providing we start with a diagram which indicates clearly which is the actual velocity along the path we travel and which are the components that contribute to this.

It is not always easy to find clear descriptions for the velocities and it is vital to distinguish between, in this example, the velocity of the boat relative to the water and the velocity relative to the

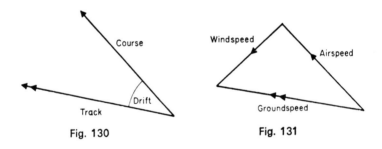

Fig. 130   Fig. 131

banks. The resultant velocity is always the velocity with which the body actually moves, whatever velocities have contributed to this motion. Since many of the problems concern aircraft, when working in two dimensions as we are now,

course is the direction in which an aircraft is heading, usually given in three digit notation, e.g. 050°,
track is the direction of the path of the aircraft over the ground,
drift is the angle from the course to the track.

The track is only the same as the course if the wind is blowing directly behind or directly against the aircraft, or if there is no wind at all. Otherwise the wind blows the aircraft off its course onto the track.

Further,

> *airspeed* is the speed with which the aircraft flies in still air,
> and the
> *groundspeed* is the speed of the aircraft over the ground.

Using vector addition,

airspeed + windspeed = groundspeed.

Note also that a north-east wind, for example, blows *from* the north-east.

Example. *An aircraft has an airspeed of 300 km h⁻¹ NE. There is a wind blowing from NNE at 50 km h⁻¹. Find the groundspeed of the aircraft and the path the aircraft makes good.*

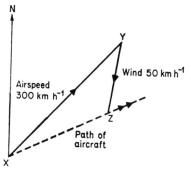

Fig. 132

The problem is illustrated in Fig. 132, which need not be drawn exactly to scale at this stage. If there were no wind, the plane would fly from X to Y. As there is a wind, the wind takes it from Y to Z. The plane actually travels along the path XZ. The lengths XY, YZ and XZ can represent velocities or they can represent distances travelled in any convenient interval of time.

Solution by scale drawing

If we draw an accurate figure we shall be able to measure XZ and the angle NXZ and hence write down the groundspeed and the direction in which the plane travels.

Take a suitable scale, say 1 cm to represent 25 km h^{-1} in this example. Draw a line up the paper to represent due north and draw XY 12 cm long in the north-east direction, i.e. angle NXY will be 45°.

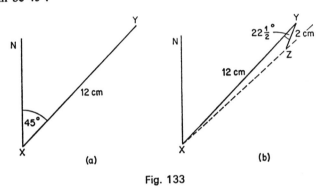

(a)   (b)

Fig. 133

Draw YZ 2 cm long *from* the NNE direction through Y, i.e. angle XYZ = 45° − 22½° = 22½°.

By measurement, XZ = 10.2 cm and angle NXZ = 49°. Hence the groundspeed of the aircraft is 25 × 10.2, i.e. 255 km h^{-1} in the direction N 49° E.

Solution by calculation

As there is not a right-angled triangle in the figure we shall need to use the cosine formula and the sine formula. Applying the cosine formula, in Fig. 133,

$$XZ^2 = 300^2 + 50^2 - 2 \times 50 \times 300 \cos 22\tfrac{1}{2}°$$
$$= 92\,500 - 30\,000 \cos 22\tfrac{1}{2}°$$
$$= 64\,780$$
$$\Rightarrow XZ = 254.5.$$

Applying the sine formula,

$$\sin YXZ = \frac{50 \sin 22\tfrac{1}{2}°}{254.5}$$

Using log sines,

$$\text{angle } YXZ = 4° 18',$$

Thus the groundspeed of the aircraft is 255 km h^{-1}, correct to three significant figures, in the direction N 49° 18' E.

The solution by drawing was far quicker.

Use of unit vectors

If the data of a problem like this one is presented in terms of unit vectors the solution is much quicker.

A plane with an airspeed of $(140\mathbf{i} + 140\mathbf{j})$ km h^{-1} meeting a wind of $(-20\mathbf{i} - 40\mathbf{j})$ km h^{-1} has a groundspeed of

$$(140\mathbf{i} + 140\mathbf{j}) + (-20\mathbf{i} - 40\mathbf{j}) = (120\mathbf{i} + 100\mathbf{j}) \text{ km h}^{-1},$$

from which the magnitude and direction of the groundspeed is obtained at once. The ease with which this can be extended into three dimensions deserves demonstration here, although not at present examined by many Boards at this level.

Example. *A glider launched with velocity $(40\mathbf{i} + 40\mathbf{j} + 20\mathbf{k})$ km h^{-1} meets an aircurrent velocity $(10\mathbf{i} - 20\mathbf{j} + 10\mathbf{k})$ km h^{-1}, $\mathbf{i}$ and $\mathbf{j}$ being unit horizontal vectors and $\mathbf{k}$ a unit vertical vector. Find the resultant velocity of the glider.*

The resultant velocity of the glider is

$$(40\mathbf{i} + 40\mathbf{j} + 20\mathbf{k}) + (10\mathbf{i} - 20\mathbf{j} + 10\mathbf{k}) \text{ km h}^{-1}$$
$$= (50\mathbf{i} + 20\mathbf{j} + 30\mathbf{k}) \text{ km h}^{-1}.$$

Applying Pythagoras theorem to find the magnitude of this vector, the resultant speed of the glider is

$$\sqrt{(50^2 + 20^2 + 30^2)} = \sqrt{3800}$$
$$= 61.6 \text{ km h}^{-1}.$$

Resultant motion in a specified direction

In these examples we have supposed that the oarsman and the pilot headed in a certain direction and we have found where the current or the wind carried them. It happens more frequently that they know where they wish to go and want to determine which course will take them to their destination.

Suppose that in the first case we want to row across the river from a point A to the point B directly opposite A. We must set such a course that the resultant velocity of the boat is along AB. Fig. 135 illustrates such a course and gives us the data to solve the problem.

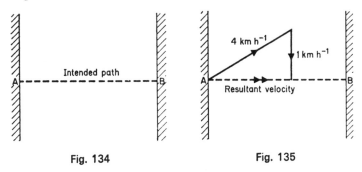

Fig. 134   Fig. 135

From the figure we see we must make an angle θ with AB, where

$$\sin \theta = \tfrac{1}{4}$$
$$\Rightarrow \theta \simeq 14\tfrac{1}{2}°,$$

and that our speed across the river will be $\sqrt{15}$ km h^{-1}, i.e. 3.87 km h^{-1} correct to three significant figures. Thus if the river is 100 m wide it will take us about 0.258 h, i.e. $15\tfrac{1}{2}$ minutes, to cross the river. If we aimed directly across, as in Fig. 134, we should cross in $\tfrac{1}{4}$ h, i.e. 15 minutes. But then, of course, we should not land at B.

Harder example

Consider again the aircraft that we met first on p. 173. Suppose the pilot has to fly from an airport A to another airport B, 400 km north-east of A. We have found that if he heads north-east he will be blown onto a course 4 °from the one he wished to follow. The pilot knows the path he wants to take, he knows the magnitude and direction of the velocity of the wind and he knows the magnitude of his airspeed. In what direction should he head to reach B? The data are illustrated in Fig. 136.

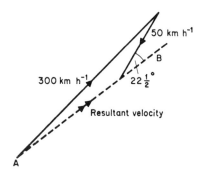

Fig. 136

Solution by drawing

Using the same scale as before, 1 cm to 25 km h⁻¹ draw a line across the paper to represent the path AB. Take any convenient point on the path, which may as well be B, and draw the line BC 2 cm in length at an angle of $22\frac{1}{2}°$ with AB produced, to represent the speed of the wind.

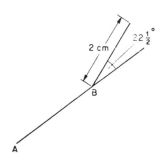

Fig. 137

The line which represents the airspeed must be 12 cm long and must finish at C, so with compass point at C draw an arc radius 12 cm to cut AB (produced if necessary) at D. Then DC will represent the airspeed.

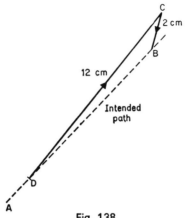

Fig. 138

From Fig. 138, DB = 9.8 cm. Hence the groundspeed is 245 km h^{-1}. Angle CDB = 5°. Hence the direction in which the pilot must head is N 40° E.

Notice that there are two points in which the arc centre C radius 12 cm could cut AB produced. The second point, D′,

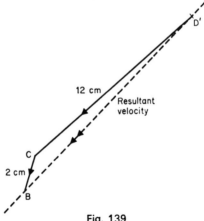

Fig. 139

gives the groundspeed for the return journey. These two will not be confused if the velocities are clearly indicated by arrows, as in these figures.

From Fig. 139, D′B = 13.6 cm, hence the groundspeed is 340 km h⁻¹.

Angle BD′C = 5°, hence the course the pilot must set for home is S 50° W.

Time of flight

Since the groundspeed from A to B is 245 km h⁻¹ and the distance is 400 km, the journey from A to B takes 1.635 h.

Since the groundspeed from B to A is 340 km h⁻¹, the journey from B to A takes

$$400/340 \simeq 1.176 \text{ h.}$$

The total journey therefore will occupy about 2.81 h, i.e. 2 hours 49 minutes.

Solution by calculation

This is longer than the solution by drawing but is necessary if the problem is algebraic, or if greater accuracy is required.

Referring to Fig. 136, let the groundspeed of the aircraft from A to B be x km h⁻¹. Then by the cosine rule,

$$300^2 = 50^2 + x^2 - 2(50)x \cos 157\tfrac{1}{2}°$$

$$\Rightarrow x^2 + 92.39x - 87\,500 = 0$$

by the formula for solving a quadratic equation,

$$x = 253 \text{ or } -346,$$

the groundspeed from A to B is 253 km h⁻¹.

The negative value gives the speed for the return journey. (It will be seen to be obtainable from triangle BD′C, since cos D′CB = − cos DCB, Fig. 139.)

For angle CDB, the sine formula gives

$$\sin \text{CDB} = \frac{50 \sin 22\tfrac{1}{2}°}{300}.$$

Hence, as before,   CDB = 4°.

EXERCISE 21

1. An aircraft on a course of 090°, airspeed 250 km h⁻¹ experiences a wind of 25 km h⁻¹ from 025°. Find its track and groundspeed.

2. An aircraft flying at an airspeed of 260 km h⁻¹ on a course of 290° encounters a wind of 25 km h⁻¹ from 340°. Find its track and groundspeed.

3. In an aircraft flying at an airspeed of 200 km h⁻¹, the pilot wishes to travel on a track of 080°. If the windspeed is 40 km h⁻¹ from 300°, what will be the course and the groundspeed?

4. A navigator knows that his course is 090° and his airspeed is 275 km h⁻¹. If the wind is 20 km h⁻¹ from 160°, find the track and groundspeed.

5. The airspeed of an aircraft is 240 km h⁻¹ and the course set is 080°. A wind of 40 km h⁻¹ is blowing from 320°. Find the groundspeed of the aircraft and the angle of drift.

6. An aircraft whose airspeed is 220 km h⁻¹ wishes to make good a track of 040°. The windspeed is 30 km h⁻¹ from 310°. Find the course and the groundspeed.

7. An aircraft flies from an airport P to another airport Q, 400 km due north of P. The airspeed is 350 km h⁻¹ and the wind is blowing from the east at 40 km h⁻¹; find the course that the pilot sets.

8. The pilot of an aircraft, airspeed 250 km h⁻¹, wishes to fly from an airport P to another airport Q, Q being 400 km due east of P. The wind speed is 40 km h⁻¹, and the wind is coming from due north. Find the course the pilot must set and how long the journey takes. Without doing any detailed calculations, write down the course he must set for the return flight and the time that the return journey takes.

9. An airport X is 500 km from an airport Y, the bearing of Y from X being 040°. The pilot of an aircraft, airspeed 300 km h⁻¹, wishes to fly from X to Y. There is a wind of 30 km h⁻¹ from 310°. Find the course the pilot sets, the angle of drift, and the time taken for the flight from X to Y. How long will the return flight take, if the velocity of the wind is unaltered?

10. The bearing of a port P from a ship at the point Q is 230°, and the distance PQ is 5.4 km. There is a current of 5 km h⁻¹ from 320°. Find the course a ship must steer to go from Q to P, if the ship can maintain a steady speed of 12 km h⁻¹ in still water. How long will it take to go from Q to P?

11. The pilot of an aircraft, airspeed 250 km h⁻¹, wishes to fly from an airport P to another airport Q, Q being 400 km due east of P.

The wind speed is 40 km h⁻¹ from 045°. Find the course the pilot sets and the time for the flight from P to Q.

If the velocity of the wind is unaltered, what course must the pilot set for the return flight from Q to P, and how long will the flight take?

12. A man wishes to row from a point P on one bank of a river to a point Q on the opposite bank, 32 m upstream from P. The river is 80 m broad. If the speed of the current is 2 km h⁻¹ and the man can row at 5 km h⁻¹, how long will it take him to row from P to Q? How long will the return journey take?

13. A man wishes to row from a point P on one bank of a river to a point Q on the opposite bank, 40 m downstream. The river is 160 m broad. If the speed of the current is 1 km h⁻¹ and the man can row at 5 km h⁻¹ how long will the journey take? How long will it take him to row back from Q to P?

14. An intelligent pigeon is 20 km south of its home. The pigeon reckons it can maintain a speed of 15 km h⁻¹, and knows there is a wind of 8 km h⁻¹ from 040°. What course should the pigeon set to reach its home, and how long will the journey take?

15. *Solution by drawing recommended*

A ship wishes to sail in a north-easterly direction but must not go within 100 m of a buoy, 500 m distant on a bearing of 050° from the present position of the ship. Within what limits must the course of the ship *not* lie if it is never within 100 m of the buoy?

Subtraction of velocities: apparent velocity

It is always slightly exciting to be in an express train when it passes near another express travelling in the opposite direction. If each train is travelling at 150 km h⁻¹, the second train appears to a man in the first train to have a speed of 300 km h⁻¹, the vector difference of the velocities of the trains. Likewise, when an express train travelling at 150 km h⁻¹ overtakes a slow train travelling at 60 km h⁻¹, the slow train has an apparent velocity of 90 km h⁻¹ backwards relative to a passenger in the express. Remember how the slow train appeared to fall away behind you?

The *apparent velocity* of a body B to an observer A is the velocity of B — the velocity of A.

Although our examples illustrate cases where the bodies are moving in the same direction, the definition of apparent velocity is still applicable in whatever direction the body and the observer are moving. The 'difference' is found by vector subtraction.

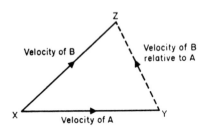

Fig. 140

Notice the check of signs. In the triangle XYZ,

$$\overrightarrow{XZ} = \overrightarrow{XY} + \overrightarrow{YZ}$$

since

velocity of B = velocity of A + (velocity of B − velocity of A).

If the velocities are given in vector form, the subtraction is easy. Otherwise the methods which we used earlier in the chapter must be adapted. Again, it is usually easier to solve the vector triangles by drawing than by calculation.

Example 1. *The velocity of a particle A is $(3\mathbf{i} + 4\mathbf{j})$ m s^{-1}. The velocity of a particle B is $(\mathbf{i} - 5\mathbf{j})$ m s^{-1}. What is the velocity of B relative to A?*

Velocity of B relative to A = velocity of B − velocity of A

$$= (\mathbf{i} - 5\mathbf{j}) - (3\mathbf{i} + 4\mathbf{j})$$

$$= (-2\mathbf{i} - 9\mathbf{j}) \text{ m s}^{-1}.$$

Notice that the velocity of A relative to B is $(2\mathbf{i} + 9\mathbf{j})$ m s^{-1}, equal in magnitude but opposite in direction to the velocity of B relative to A.

Example 2. *The particle A is moving along a straight line with velocity 3 m s⁻¹. The particle B has velocity 5 m s⁻¹ at an angle of 30° to the path of A. Find the velocity of B relative to A.*

Solution by drawing

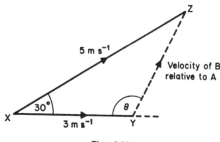

Fig. 141

Take a scale of 2 cm to represent 1 m s⁻¹.
Draw a line XY 6 cm long to represent the velocity of A.
Draw a line XZ 10 cm long to represent the velocity of B.
Then YZ represents the velocity of B relative to A.
From an accurate drawing ZY = 5.6 cm and angle $\theta = 118°$ hence the velocity of B relative to A is 2.8 m s⁻¹ at 62° with the path of A.
Notice that as always

velocity of B = velocity of A + (velocity of B − velocity of A),

i.e. $\overrightarrow{XZ} = \overrightarrow{XY} + \overrightarrow{YZ}$.

Solution by calculation
From the cosine formula,

$$v^2 = 5^2 + 3^2 - 2 \times 5 \times 3 \times \cos 30°,$$

where v m s^{-1} is the relative velocity required. Hence

$$v^2 = 34 - 25.98$$

$$= 8.02$$

$$\Rightarrow v = 2.83.$$

From the sine formula,

$$\frac{5}{\sin \theta} = \frac{2.832}{\sin 30°}$$

$$\Rightarrow \sin \theta = \frac{5 \sin 30°}{2.832}$$

$$\Rightarrow \theta = 62° \text{ or } 118°$$

Since 5 cos 30° is greater than 3, the projection of XZ on XY is greater than XY, so the obtuse angle is required. Care must always be used when this ambiguity occurs.

Problems in which we do not know both velocities but have some information about their difference can also be solved. These require careful thought and a clear diagram.

Example *To a man walking due west at 5 km h^{-1} the wind appears to come from S 22½° W. When he walks due south at the same speed, the wind appears to come from S 22½° E. Find the speed and true direction of the wind, assumed constant throughout.*

Solution by drawing

Since the velocity of the wind is constant, it must be represented by the same vector throughout the problem. Using the first set of data, we can begin to build up a vector diagram, taking 2 cm to represent 1 km h^{-1}, as in Fig. 142.

The dotted line – – – – – which represents the velocity of the wind relative to the man is of unknown length, as we do not know the relative velocity. Adding the second set of data, Fig. 143,

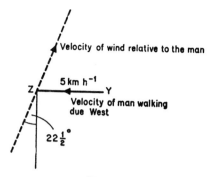

Velocity of wind relative to the man

5 km h⁻¹

Velocity of man walking
due West

$22\frac{1}{2}°$

Fig. 142

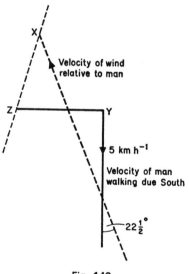

Velocity of wind
relative to man

5 km h⁻¹

Velocity of man
walking due South

$22\frac{1}{2}°$

Fig. 143

we can now complete the figure with the vector representing
the velocity of the wind, in Fig. 144.

The velocity of the wind is 5 km h⁻¹, since XY = 10 cm, and
the true direction of the wind is from the south-east, since
angle XYZ = 45.°

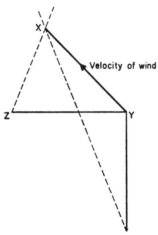

Fig. 144

Solution by calculation

Denoting the velocity of the wind by v km h^{-1}, at an angle θ° to the initial path of the man,

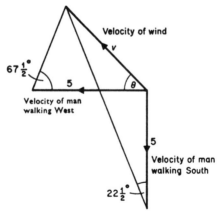

Fig. 145

By the sine formula,

$$\frac{v}{\sin 22\frac{1}{2}^{\circ}} = \frac{5}{\sin (67\frac{1}{2} - \theta)^{\circ}}$$

and
$$\frac{v}{\sin 67\frac{1}{2}°} = \frac{5}{\sin (112\frac{1}{2} - \theta)°}$$

$$\Rightarrow \frac{\sin 22\frac{1}{2}°}{\sin 67\frac{1}{2}°} = \frac{\sin (67\frac{1}{2} - \theta)°}{\sin (112\frac{1}{2} - \theta)°}$$

$$\Rightarrow \sin 22\frac{1}{2}° \sin (112\frac{1}{2} - \theta)° = \sin 67\frac{1}{2}° \sin (67\frac{1}{2} - \theta)°$$

Using the trigonometrical identity

$$2 \sin A \sin B = \cos(A - B) - \cos(A + B)*$$

we have

$$\cos(\theta - 90)° - \cos(135 - \theta)° = \cos \theta° - \cos(135 - \theta)°$$

Hence $\theta = 45°$ is the only solution between 0° and 90°.

Substituting in the first equation gives $v = 5$, or we can look at the geometry of Fig. 145 and see that we have two isosceles triangles. Thus we also obtain the result that the wind speed is 5 km h^{-1}, from the south east.

EXERCISE 22

1. A navigator steers a course of 050° and the airspeed is 220 km h^{-1}. After 20 minutes, he notes that he has made good a track of 047° and the distance travelled is 75 km. What are the speed and direction of the wind?

2. The navigator of an aircraft flying at an airspeed of 220 km h^{-1} on a course of 135° fixes his position at 0900 hours. A second fix is obtained at 0915 hours. The distance and direction of the second fix from the first are 50 km and 125°. Find the speed and direction of the wind.

3. A ship sailing due north at 30 km h^{-1} sees a hovercraft apparently approaching from the east at 60 km h^{-1}. Find the true velocity of the hovercraft.

4. A light aircraft flying on a track of 045° at 200 km h^{-1} sees another aircraft apparently flying at 150 km h^{-1} on 070°. Find the true velocity of the other aircraft.

5. A spacecraft velocity $-32\,000\mathbf{i}$ km h^{-1} sees another spacecraft with apparent velocity $(40\,000\mathbf{i} + 6000\mathbf{j})$ km h^{-1}. Find, in vector form, the true velocity of the second spacecraft. What is the true speed of the second craft?

* See *Additional Pure Mathematics*, page 277.

6. A water-boatman can average 4 mm s^{-1} on still water. If he heads up and across the stream, yet travels downstream at 4 mm s^{-1} at 40° to the bank of the stream, what is the speed of the water? (Assume the water flows parallel to the banks of the stream.)

7. To a man walking due north at 6 km h^{-1} the wind appears to come from N 30°W. When walking due east at 6 km h^{-1}, the wind appears to come from N 30°E. What is the true velocity of the wind?

8. To a girl riding due north at 9 km h^{-1} the wind appears to come from the west. When she walks north-east at 8 km h^{-1} the wind appears to come from the south. Find the true magnitude and direction of the wind.

9. To a man sailing due south at 10 km h^{-1} the wind appears to be coming from S 30°E. When sailing at 5 km h^{-1} due west, the wind appears to be coming from S 60°E. Find the true magnitude and direction of the wind.

10. A boy swims at 3 km h^{-1} and wishes to cross a river 400 m wide. The speed of the current is 6 km h^{-1}. Find how long it will take the boy to cross if he swims in such a manner as to land as little downstream as possible.

CHAPTER 14

Projectiles

Path of a projectile

In Chapter 1 we considered the motion of a particle projected vertically upwards or vertically downwards. We are probably all familiar with the path in which a cricket ball or a stone travels when thrown, and we may know that the curve is called a parabola, from the Greek word meaning 'to throw'. The path is sometimes called the trajectory.

To look in detail at the path of a projectile, first consider a particle projected horizontally with velocity 1 m s^{-1} from a point high above the ground. It is convenient to refer the curve in which the particle travels to rectangular cartesian coordinate axes, to take the point of projection as the origin, the x-axis horizontal and the y-axis vertical.

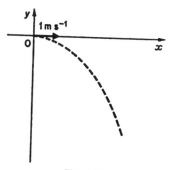

Fig. 146

Since there is no horizontal acceleration,

>after 1 s the horizontal distance travelled is 1 m,
>after 2 s the horizontal distance travelled is 2 m,
>after 3 s it is 3 m, and so on.

These give the x coordinates of the position of the particle 1, 2, 3 or more seconds after projection.

Considering the vertical motion, the acceleration is -9.8 m s^{-2} (negative since the direction in which y is increasing is upwards and gravity acts downwards), and the initial vertical velocity is zero. Using

$$s = ut + \tfrac{1}{2}at^2,$$
$$y = -4.9t^2.$$

Hence we get

after 1 s the y coordinate of the position of the particle is -4.9,
after 2 s the y coordinate is -19.6,
after 3 s it is -44.1, etc.

The positions of the particle at some points in its motion are shown in Fig. 147; check the coordinates of those not calculated above.

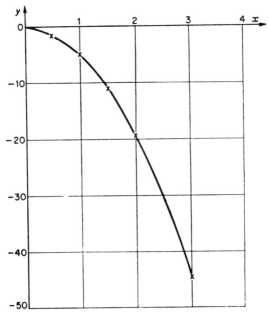

Fig. 147

Our ideas of the shape of the curve in which the particle travels should be confirmed by the sketch in Fig. 147.

More generally, if the initial velocity of the particle is u m s^{-1} horizontally and the coordinates of the particle after t seconds are (x,y), we have $x = ut$ and $y = -\frac{1}{2}gt^2$. Hence

$$y = -\frac{1}{2}g\left(\frac{x}{u}\right)^2 = -\frac{g}{2u^2}x^2$$

The reader will undoubtedly be familiar with curves like $y = 3x^2$, and will have recognized the curve in Fig. 147 as belonging to that family.

Example. *A ball is thrown horizontally with velocity 6 m s^{-1} from the top of a high building. Find how far it has fallen when it strikes a vertical wall 12 m from the point of projection.*

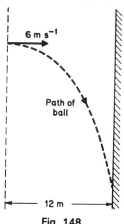

6 m s^{-1}

Path of
ball

12 m

Fig. 148

When there is a choice, always consider the horizontal motion before the vertical motion. Considering the horizontal motion of the ball, it takes 2 s to travel the 12 m to the wall.

Considering the vertical motion, with the usual notation,

$$u = 0, \qquad a = -9.8, \qquad t = 2,$$

and we want to find s, i.e.

$$s = \frac{1}{2}(-9.8)2^2 = -19.6,$$

the ball has fallen 19.6 m when it strikes the wall.

With more difficult problems it will sometimes be worth expressing these vectors in terms of their horizontal and vertical components. In this case, $u = 6\mathbf{i}$ and $a = -9.8\,\mathbf{k}$, and there is little benefit in adopting this notation.

EXERCISE 23

1. A stone is thrown horizontally from the top of a vertical cliff with velocity 10 m s^{-1} and strikes the sea 40 m from the foot of the cliff. Find the height of the cliff.

2. A boy throws himself from the top of a diving board with a horizontal velocity of 2 m s^{-1}. If he lands in the swimming pool 3 m from the point vertically below his point of projection, how high is the diving-board?

3. A cricket ball passes horizontally over a sightscreen with a velocity of 20 m s^{-1} and lands 25 m beyond the sightscreen. What was the height of the ball above the ground when it passed over the sightscreen?

4. A batsman hits a cricket ball horizontally with a velocity of 15 m s^{-1} and the ball is caught by a fielder standing 5 m from the bat. If the ball was 0.6 m above the ground when hit by the batsman, how high was it above the ground when caught by the fielder?

5. A tennis player serves a ball horizontally with velocity 25 m s^{-1} and it strikes the ground 18 m horizontally from the point vertically below the point of projection. At what height above the ground was the ball served?

Projection at an angle to the horizontal

In all these examples the bodies were projected horizontally and of course air resistance and spin were ignored. These are the reasons why a football kicked a great distance rarely moves in a parabola, and why experts in many sports are able to vary the paths in which they project the balls used. Some of these last factors are difficult to analyse mathematically, but we can easily consider bodies projected at an angle to the horizontal, ignoring the other variations.

Example *A body is projected from a point P on the ground with a velocity of 8 m s^{-1} at an angle of 30° above the horizontal. How far from P does it return to the horizontal plane through P?*

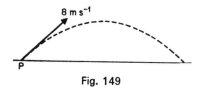

Fig. 149

Considering the vertical motion,

$$u = 8 \sin 30°$$

(the vertical component of the initial velocity is 8 sin 30° m s^{-1}).
Since $s = 0$ (when it strikes the ground again) and $a = -9.8$,
from

$$s = ut + \tfrac{1}{2}at^2$$

we have $\qquad\qquad 0 = 4t + \tfrac{1}{2}(-9.8)t^2$

i.e. $\qquad\qquad\qquad t = 0 \text{ or } \dfrac{8}{9.8}$

i.e. $t = 0$ or 0.8, to 1 decimal place. The value $t = 0$ obviously
corresponds to the time at the moment of projection, and we see
that the time of flight of the body is 0.8 s.

Considering the horizontal motion, in which there is no acceleration,

$$s = ut$$
$$\Rightarrow s = (8 \cos 30°) \times 0.8$$
$$= 5.5 \text{ to 1 decimal place.}$$

The body lands 5.5 m from the point of projection.

Symmetry of path

Returning to the vertical motion of the body in the last example,
the vertical component v m s^{-1} of the velocity with which the
body lands is given by

$$v = u + at$$
$$\Rightarrow v = 4 - (9.8) \times \left(\dfrac{8}{9.8}\right)$$
$$= -4,$$

the vertical component of the velocity with which it lands being equal in magnitude to the vertical component of the velocity with which it was projected. Since the horizontal velocity is unchanged throughout the motion, the velocity with which a body returns to the level with which it was projected is equal in both magnitude and inclination to the horizontal to the velocity of projection.

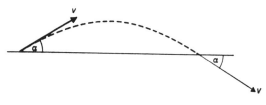

Fig. 150

This could have been anticipated from the symmetry of the figure. It is often easier to find the time of flight by making use of this property instead of finding the time until s is again 0.

Example 1. *A batsman hits a cricket ball with velocity* $(10\mathbf{i} + 14\mathbf{k}) m s^{-1}$ *and is caught by a fielder at the same height above ground as the point P where the ball was struck. Find the distance from P of the fielder.*

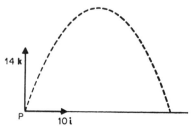

Fig. 151

The vector notation gives us the components of the initial velocity very easily. Considering the vertical motion,

$$u = 14$$
$$v = -14 \text{ (from the symmetry of the figure),}$$
$$a = -9.8,$$

and we wish to find t. From

$$v = u + at$$

we have $$-14 = 14 - 9.8t$$

i.e. $$t = \frac{28}{9.8} = 2.9 \text{ s to 1 decimal place.}$$

From the horizontal motion,

$$x = 10 \times 2.9 = 29.$$

The fielder is 29 m from the point where the ball was struck.

Example 2. *A golfer hits a ball with velocity 35 m s^{-1} onto a green 125 m away. Find the angle of projection of the ball.*

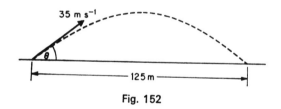

Fig. 152

Let the angle of projection be θ and the time of flight be t seconds. Then considering the horizontal motion,

$$125 = (35 \cos \theta)t. \tag{i}$$

Considering the vertical motion, and using the symmetry of the path,

$$v = u + at$$
$$\Rightarrow -35 \sin \theta = 35 \sin \theta - 9.8t$$
$$\Rightarrow 9.8t = 70 \sin \theta.$$

Substituting for t in (i),

$$125 = (35 \cos \theta) \times \frac{70 \sin \theta}{9.8}$$

$$= 250 \sin \theta \cos \theta$$

Hence
$$1 = 2 \sin \theta \cos \theta$$

$$\Rightarrow \sin 2\theta = 1*$$

$$\Rightarrow 2\theta = 90°$$

$$\Rightarrow \theta = 45°,$$

the golfer must hit the ball at 45° above the horizontal.

Time of flight: range on a horizontal plane through the point of projection

Many of the results we have found can be obtained algebraically, taking the velocity of projection as V and the angle of projection as θ above the horizontal (θ can of course be negative).

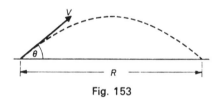

Fig. 153

Considering the vertical motion, $u = V \sin \theta$, $a = -g$, and $s = 0$ when the body returns to the horizontal plane through the point of projection.

$$s = ut + \tfrac{1}{2}at^2$$

$$\Rightarrow 0 = V \sin \theta \, t - \tfrac{1}{2}gt^2$$

$$\Rightarrow t = 0 \text{ or } \frac{2V \sin \theta}{g},$$

the value $t = 0$ occurring as in the first example. The time of flight is therefore $2V \sin \theta / g$.

*Using the identity $\sin 2\theta = 2 \sin \theta \cos \theta$. See Chapter 26 of *Additional Pure Mathematics*.

To find the range R on the horizontal plane, using $s = ut$ we have

$$R = (V \cos \theta) \frac{2V \sin \theta}{g} = \frac{V^2 \sin 2\theta}{g},$$

using the trigonometric identity as on the previous page.

To find the velocity with which it returns to the plane, from the vertical motion,

$$v = u + at$$

$$\Rightarrow v = V \sin \theta - g \frac{(2V \sin \theta)}{g}$$

$$= -V \sin \theta,$$

as before, so the speed with which the body returns to the level of projection is always equal to the speed of projection.

Maximum range

Since the range on a horizontal plane is $V^2 \sin 2\theta/g$, the greatest value of the range will occur when $\sin 2\theta = 1$, i.e.

$$2\theta = 90°$$

or

$$\theta = 45°.$$

The maximum range which can be obtained with a given velocity of projection V is V^2/g and this is obtained when the angle of projection is 45°.

EXERCISE 24

1. A boy throws a stone from a point P with velocity 28 m s^{-1} at 30° above the horizontal. Find
    (i) the time before the stone reaches the highest point X in its path,
    (ii) the height of X above the horizontal plane through P,
    (iii) the time which elapses after projection before the stone returns to the horizontal plane through P,
    (iv) the range on the horizontal plane.

2. A shell is fired from a point P at 490 m s^{-1} at 45° above the horizontal. Find
   (i) the greatest height reached by the shell,
   (ii) the range on the horizontal plane through P.

3. A cricketer hits a ball with velocity $(21\mathbf{i} + 14\mathbf{k})$ m s^{-1}. How far from the bat does the ball return to the horizontal level of the bat?

4. A tennis ball is served from a point 2.5 m above the ground with a velocity of $(26\mathbf{i} - 0.5\mathbf{k})$ m s^{-1}. Find the time that elapses before it is above the net, and how high it is above the net when it passes over the net, assumed 1 m high and 13 m horizontally from the point at which the ball was served.

5. A cricketer hits a ball with velocity 25 m s^{-1} at 60° above the horizontal. Find how far above the ground it passes over a fielder 50 m from the bat. (Assume the ball is struck very close to the ground.)

6. A cricketer hits a ball at 30 m s^{-1} at 45° above the horizontal. He scores 6 runs if it travels over the boundary, 60 m from the bat, before it hits the ground. Does he score 6 runs for this hit?

In questions 7—11, α is the angle given by cos α = $\frac{4}{5}$, sin α = $\frac{3}{5}$.

7. A boy catapults a stone at 19.6 m s^{-1} at α above the horizontal. Find the greatest height reached by the stone, and its range on the horizontal plane through the point of projection.

8. A football is kicked with velocity 14.7 m s^{-1} at α above the horizontal. Find the greatest height reached by the football and its range on the horizontal plane through the point of projection.

9. Two stones are projected from a point P in the same direction, with velocity 20 m s^{-1}, one at α above the horizontal and the other at α to the vertical. Find the distance of each stone from P 1 s after projection, and the distance the stones are apart.

10. A shell is fired from a gun with muzzle velocity 250 m s^{-1} at α above the horizontal. Find its height above the target, 6 km from the gun and in the same horizontal plane as the gun.

11. A shell is fired from a gun at 150 m s^{-1} at α above the horizontal. Find the height of the shell above its target, 1800 m horizontally from the gun and 240 m above the level of the gun.

12. Two stones are thrown simultaneously from a point at the top of a cliff, one at θ above the horizontal and the other at θ below the horizontal. Show that one stone will always be vertically above the other.

13. An aircraft flying horizontally at 360 km h^{-1} releases a bomb (with no velocity relative to the aircraft) at a stationary tank 200 m away. What must be the height of the aircraft above the tank if the bomb is to hit the tank?

14. A man throws a dart with a velocity of 10 m s^{-1} at an angle θ to the horizontal. The dart strikes the board 7.5 m from his hand, and at the same horizontal level as the hand. Find the two possible values of θ.

15. An aircraft flying horizontally at 360 km h^{-1} drops a bomb from a height of 1960 m when vertically above the target. How far from the target does the bomb land?

16. A bird flying at 10 m s^{-1} at 30° above the horizontal drops a breadcrumb. If the bird is 5 m above the ground when it drops the breadcrumb, how far above the ground is the crumb 0.4 s after it was dropped?

17. A golfer hits a ball at 30° above the horizontal and it lands on a green, at the same level as the tee, 100 m from the tee. Find the velocity with which it was struck.

18. A polo player strikes a ball at 15° to the horizontal. The ball lands on the level field 40 m from the point where it was struck. Find the velocity given to the ball when hit.

19. A tennis player 'smashes' a ball from a height of 2.8 m, so that it leaves the racket with velocity 37.8 m s^{-1} at 30° below the horizontal. How far from the feet of the player does the ball hit the ground? (Assume his feet are vertically below the point at which he struck the ball.)

20. A golfer hits the ball from a tee with velocity 49 m s^{-1}, and sees it land in the 'rough' 196 m away, 24.5 m above the level of the tee. If t s is the time of flight of the golf ball and $\theta°$ the angle its initial velocity makes with the horizontal, form two equations in t and θ, and hence find the angle above the horizontal at which the ball was struck.

CHAPTER 15

Machines

A machine is any device whereby a force can be applied in a more convenient manner. It may be that we wish to raise a car with a punctured tyre. We cannot lift the car, but a jack enables us to apply a force, small compared with the weight of the car, over a distance which is much greater than the height through which we wish to raise the car.

Effort: load

The force that is applied to the machine is called the *effort*: the resisting force it overcomes is called the *load*. The usual object of a machine is to enable a relatively small effort to overcome a large load. This is not always the case; a machine may be used to apply a force at a point inaccessible to the effort. Consider the dentist's forceps, which apply a considerable force at an inaccessible point, or the delicate instruments used in surgery which apply a small force at an inaccessible point. Work is never saved by a machine. As a perfectly efficient machine is almost impossible, work in fact is normally lost by a machine.

Mechanical advantage (or force ratio), velocity ratio, efficiency.

The *mechanical advantage*, sometimes called the force ratio, of a machine compares the load L and effort E.

$$\text{Mechanical advantage} = \frac{L}{E}.$$

Thus the mechanical advantage is usually greater than 1.

The *velocity ratio* is the ratio of the velocity of the point of application of the effort E to the velocity of the point of application of the load L. Since it is usually easier to measure or calculate distances than velocities,

$$\text{Velocity ratio} = \frac{\text{distance moved by the point of application of } E}{\text{distance moved by the point of application of } L}.$$

Thus the velocity ratio is usually greater than 1.

The efficiency of a machine compares the work done by the effort with the work used overcoming the load.

$$\text{Efficiency} = \frac{\text{work done by load } L}{\text{work done on effort } E}$$

$$= \frac{L \times (\text{distance moved by point of application of } L)}{E \times (\text{distance moved by point of application of } E)}$$

$$= \left(\frac{L}{E}\right)\left[\frac{\text{distance moved by point of application of } L}{\text{distance moved by point of application of } E}\right]$$

$$= \frac{\text{mechanical advantage}}{\text{velocity ratio}}.$$

If there is not any friction and unnecessary work is not done raising pulleys, the efficiency is equal to 1.

Levers

Some of the commonest examples of machines are so simple that we do not think of them as machines. The lever has many applications—bottle openers, can openers, tyre levers, spanners and in nearly every type of internal combustion engine, to name only a few. We may not be as ambitious as Archimedes is alleged to have been when he first discovered the principle, 'give me a fixed fulcrum and I can move the earth', but we use levers in various forms all the time. The fixed point X in Fig. 154 is the fulcrum.

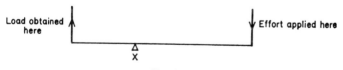

Fig. 154

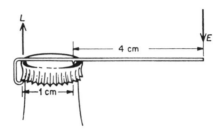

Fig. 155

A simple type of bottle opener is shown in Fig. 155. The work done by the effort E when the lever turns through an angle θ is $E \times 4\theta$. The work done against the resistance L is $L \times \theta$. Hence

$$4E\theta = L\theta$$

$$\Rightarrow 4E = L.$$

This bottle opener enables a force four times the effort to be applied to the rim of the cap of the bottle: a greater force is applied at an inaccessible point. In this example,

$$\text{Mechanical advantage} = 4,$$

$$\text{Velocity ratio} = 4,$$

$$\text{Efficiency} = 1.$$

Inclined plane

In several earlier questions we have considered the problem of raising a body up an inclined plane. We often use this device when we wish to raise vertically a body that we are unable to lift, a garden roller or a wheelbarrow, or to place a heavy load onto a lorry. In these examples rolling is used, and it is likely that the upper stones at Stonehenge, which have a mass of about 25 000 kg and are about 4 m above the ground, were put into position by building 'inclined planes'. For the size of the bodies concerned, dragging would be a good approximation to the rolling on rough wooden branches which may have been used.

Consider a body of mass m being pulled at a steady speed up a plane inclined at an angle α to the horizontal, as in Fig. 156. The effort E is raising a load mg.

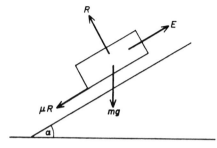

Fig. 156

Resolving perpendicular to the plane,

$$R = mg \cos \alpha.$$

Resolving along the plane,

$$E = mg \sin \alpha + \mu R$$
$$\therefore \qquad E = mg(\sin \alpha + \mu \cos \alpha).$$

Hence

$$\text{Mechanical advantage} = \frac{mg}{E} = \frac{1}{\sin \alpha + \mu \cos \alpha}.$$

$$\text{Velocity ratio} = \frac{1}{\sin \alpha},$$

$$\text{Efficiency} = \frac{\text{mechanical advantage}}{\text{velocity ratio}}$$

$$= \frac{\sin \alpha}{\sin \alpha + \mu \cos \alpha}.$$

Notice that if the plane is smooth, $\mu = 0$ and the efficiency is 1.

Other machines do not occur so frequently in everyday life, but it is most important to have experience of these. Time spent in a well-equipped mathematical or mechanical laboratory is very valuable, but it is even more valuable to observe machines in use, and several examples of these are given in the text.

Pulleys, block and tackle

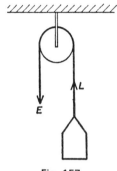

Fig. 157

Builders may often be seen raising material to the top of a building by means of a simple pulley. When the point of application of the effort moves 1 metre, the load is raised 1 metre. If the pulley is smooth, which is most unlikely,

$$\text{Mechanical advantage} = 1,$$

$$\text{Velocity ratio} = 1,$$

$$\text{Efficiency} = 1.$$

As the bearings of the axle of the pulley are not usually smooth, the efficiency will be less than 1 and the mechanical advantage less than 1. The object of this machine is to enable a downwards force to raise a body.

A much more sophisticated version is the block and tackle shown in Fig. 158. If the pulleys are smooth, the tension in the rope is always equal to E. When the point of application of E moves 1 metre, the load is raised $\frac{1}{8}$ m. (More generally, $1/n$ metres if there are n pulleys.)

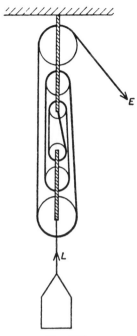

Fig. 158

The velocity ratio = 6, and if the weight of the lower block is w,

$$1E = (L + w)\frac{1}{6}$$

Since the mechanical advantage $= \dfrac{L}{E}$,

$$\text{Efficiency} = \frac{L}{E} \times \frac{1}{6}$$

$$= \frac{L}{L + w}.$$

Wheel and axle

This consists of a wheel and axle which rotate together about a common axis. The effort E is applied to a string wrapped round the wheel, and the load L is attached to a string wrapped around the axle in such a manner that when the effort is applied the load is raised.

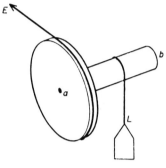

Fig. 159

If the radius of the wheel is a and the radius of the axle is b, then in one revolution the point of application of E moves $2\pi a$ and the point of application of L moves $2\pi b$.

$$\therefore \qquad \text{Mechanical advantage} = \frac{L}{E},$$

$$\text{Velocity ratio} = \frac{2\pi a}{2\pi b} = \frac{a}{b},$$

$$\text{Efficiency} = \frac{Lb}{Ea}.$$

Differential wheel and axle

The ratio effort : load is determined by the ratio of the radii of the wheel and axle, in the last example. In practice, a may have to be less than a certain value to fit into available space and to avoid other problems, and b may have to be greater than a certain value so that the axle does not break. To improve the mechanical advantage of the system the differential wheel and axle was developed.

The axle consists of two parts with different radii. The load is attached to a pulley, round which passes a continuous string wrapped in opposite directions on the two parts of the axle. An effort E on the wheel winds up the string on the larger axle and unwinds the string on the smaller axle.

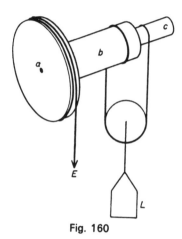

Fig. 160

If the radius of the wheel is a, and the radii of the two parts of the axle are b and c, in one revolution E moves a distance $2\pi a$ and L rises a distance $\pi(b - c)$, since the length of string pulled up is $2\pi(b - c)$.

$$\text{Velocity ratio} = \frac{2a}{b - c},$$

$$\text{Efficiency} = \frac{E}{L}\left(\frac{b - c}{2a}\right).$$

The velocity ratio can be made as large as we wish by making the difference $(b - c)$ as small as necessary.

Weston differential pulley

The upper pulley in Fig. 161 has two grooves, one having a slightly larger radius than the other. The load is attached to another pulley. An endless chain passes round the larger groove, back round the smaller groove and then round the lower pulley.

If the radii of the larger and smaller grooves are R and r, in one revolution a length $2\pi R$ of chain is drawn up by the outer groove and a length $2\pi r$ is let down by the inner groove. The

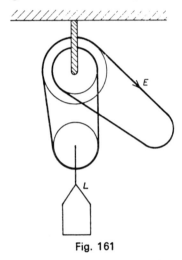

Fig. 161

distance the load is raised is half the difference of these, i.e. $\pi(R - r)$, when the point of application of E moves through a distance $2\pi R$.

$$\text{Velocity ratio} = \frac{2\pi R}{\pi (R - r)}$$

$$= \frac{2R}{R - r},$$

$$\text{Efficiency} = \frac{L (R - r)}{2ER}.$$

Again, the velocity ratio can be made as small as we wish to allow the raising of heavy loads. A further advantage is that the efficiency is often less than 50 per cent because of friction, thus preventing the load running back when the effort is removed. This running back is called overhauling.

Examples of these two types of differential pulleys can be seen in large garages, in engineering works and in steelworks.

Screw

The pitch of a screw is the distance the screwhead moves forward in one complete turn of the screw. Thus if the effort E is applied to the end of an arm of length a (e.g. a spanner) connected to a screw of pitch p which encounters a resistance L,

$$\text{Work done by the effort} = E. \, 2\pi a$$
$$\text{Work done on the load} = L. \, p.$$

Hence
$$2\pi aE = p \, L.$$

Winch

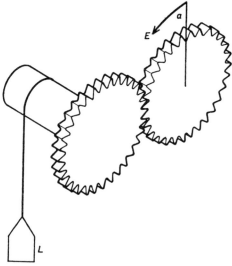

Fig. 162

The winch is a refinement of the wheel and axle. The effort E is applied at the end of an arm of length a rigidly connected to a wheel. This is cogged to a second wheel, which need not be in the same plane as the first, and is fixed to an axle radius r. If the first wheel has n cogs and the second has N cogs, in one revolution of the arm the axle will complete n/N of a revolution. The point of application of the effort E moves a distance $2\pi a$ and the point of application of the load L moves a distance $n/N \, 2\pi r$. Thus

$$\text{Velocity ratio} = \frac{2\pi a}{(n/N) \, 2\pi r}$$

$$= \frac{Na}{nr},$$

$$\text{Efficiency} = \frac{Lnr}{ENa}.$$

Winches are often seen in harbours, although so many are now electrically driven and hidden in smart casing that there is little to see.

Rack and pinion

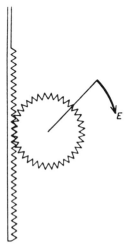

Fig. 163

This is a variation of the winch. In one revolution of the arm to which E is applied the same number of teeth on the rack are engaged as the number of teeth on the pinion, and the rack moves accordingly. If n teeth are engaged in one revolution of the pinion and the distance between each tooth is a, the rack moves a distance na. If the length of the arm is b, and a load L is attached to the rack,

$$\text{Velocity ratio} = \frac{2\pi b}{na},$$

$$\text{Efficiency} = \frac{naL}{2\pi bE}.$$

The rack and pinion is often used for raising sluice gates, but the most interesting rack and pinion is perhaps mythical. The story goes that the ruler of Syracuse in the third century before Christ had built a ship so large that it could not be launched. He wisely appealed to Archimedes for help, and Archimedes is supposed to have devised a machine which was basically a series of racks and pinions.

EXERCISE 25

1. The fulcrum of a lever AB 12 cm long is 10 cm from the end A. Find the load in newtons which can be overcome at B by a force of 10 N applied at A.
2. The oar of a boat may be regarded as a lever 3.6 m long, the rowlock serving as a fulcrum 1.2 m from the end A. If a force of 1000 N is applied at A, what force is applied by the other end of the oar?
3. In a pair of nutcrackers a nut is placed 1 cm from the hinge. The force applied by the hands may be regarded as a single force of 40 N acting at a point 10 cm from the hinge. What is the force on the nut?
4. A body of mass 500 kg is drawn up a plane inclined at 1 in 10 by a force of 600 N. What is the efficiency of the system?
5. A body is pulled up a plane inclined at 30° to the horizontal by a force parallel to the plane. If the coefficient of friction between the body and the plane is 0.4, find the efficiency of the system.
6. The relation between the effort E and the load L in a certain machine is $E = a + bL$, where a and b are constants. In this machine the velocity ratio is 15 : 1. An effort of 30 N will raise a load of 130 N and an effort of 40 N will raise a load of 200 N. Find the load that an effort of 80 N will raise, and the efficiency under this load.

7. In an experiment to determine the mechanical advantage of a machine the following readings of effort (E newtons) and load (L newtons) were obtained:

E	2	4	6	8	10
L	5.5	11.1	16.8	23.0	28.1

Plot these values to estimate the mechanical advantage of this machine.

8. In an experiment to relate the load (L newtons) overcome by an effort (E newtons) applied to a certain machine the following readings were obtained:

E	10	20	30	40	50
L	18	44	69	93	120

Plot these values and hence estimate the values of m and c if the effort and load are connected by the equation $L = mE + c$. Under what conditions is this more likely to be the relation between L and E than the relation in question 7?

9. A block and tackle has three pulleys in each block. If an effort of 40 N is necessary to overcome a load of 140 N, find the efficiency of the machine.

10. A block and tackle has two pulleys in each block, and the efficiency of the machine is 0.4. Find the load which can just be overcome by an effort of 50 N.

11. For a certain wheel and axle, the radius of the wheel can lie between 8 cm and 10 cm and the radius of the axle between 2 cm and 3 cm. Find the greatest and least values possible of the velocity ratio of this wheel and axle.

12. In a Weston differential pulley the radius of the larger groove is 10 cm and of the smaller groove is 9 cm. Find the velocity ratio and the effort needed to raise a load of 500 N, if the efficiency of the machine is 0.45.

13. In a Weston differential pulley the radius of the larger groove is 10 cm. Find the radius of the smaller groove if the velocity ratio is 25.

14. A screw with pitch 0.5 cm encounters a resistance of 100 N. What effort must be applied to the end of an arm 10 cm long to turn the screw through one complete revolution?

15. A load is to be raised by a rope passing round an axle radius 60 cm. The axle is attached to a wheel with 80 teeth geared to another wheel with 25 teeth. The effort is applied at the end of an arm length 150 cm. Find the velocity ratio.

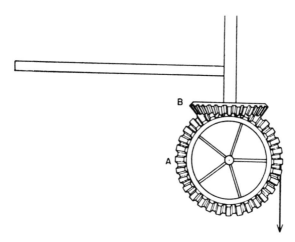

Fig. 164

16. A load is raised by a rope wrapped round a vertical wheel of radius 50 cm. The wheel is rigidly fixed to a cogwheel A with 80 teeth, which is geared to a horizontal cogwheel B with 20 teeth, as in Fig. 164. The effort is applied to the end of an arm 100 cm long which rotates B in a horizontal plane. Find the velocity ratio of the machine.

17. A press is worked by a rack and pinion. The cogs of the pinion are 1 cm apart and there are 8 cogs on the pinion. If the efficiency of the machine is 75 per cent and the crank handle is 100 cm long, what load can be overcome by an effort of 120 N?

18. A jack is worked by a rack and pinion. The cogs are 1 cm apart and there are 16 cogs on the wheel. The crank handle is 45 cm long and turns the wheel about a fixed point. What thrust is produced by an effort of 120 N if the efficiency is 35 per cent? (Take π to be $\frac{22}{7}$.)

(O. & C.*)

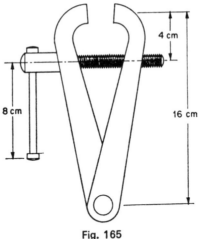

Fig. 165

19. A vice is constructed as shown in Fig. 165. The screw advances 1 cm for each revolution and the efficiency is 60 per cent. Find the force that is exerted at the middle of the jaws when a force of 280 N is applied at the end of the lever. (O. & C.*)

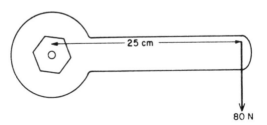

Fig. 166

20. A hexagonal nut on a machine is tightened by applying a single force of 80 N at the end of a spanner. If the pitch of the screw is 0.1 cm and the efficiency is 30 per cent, find the thrust of the nut in the direction of its axis. (O. & C.*)

CHAPTER 16

Motion in a Circle

Velocity: angular velocity

When a particle is describing a fixed circle the velocity at any moment is along the tangent at the point where the particle then is.

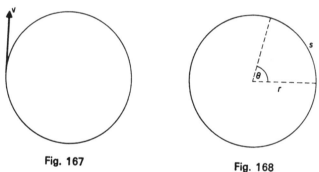

Fig. 167

Fig. 168

If the particle has constant speed v, the distance s which it travels in time t will be given by

$$s = vt$$

and the arc traversed will subtend an angle θ at the centre of the circle radius r, where

$$s = r\theta.$$

Hence

$$vt = r\theta.$$

Differentiating,

$$v = r\frac{d\theta}{dt}$$

$$\Rightarrow \frac{d\theta}{dt} = \frac{v}{r}.$$

The quantity $d\theta/dt$, which is the rate of change of the angle θ with respect to time, is called the angular velocity of the particle. This is usually denoted by ω and measured in radians per second (rad s^{-1}) in SI units. In practice, revolutions per minute (r.p.m.) or per second are more easily measured than radians, so angular velocity will often be given in terms of these and will have to be converted to radians per second.

$$1 \text{ revolution} = 2\pi \text{ radians}$$

$$\Rightarrow 1 \text{ revolution per minute} = 2\pi \text{ radians per minute}$$

$$= \frac{2\pi}{60} \text{ radians per second}$$

$$= \frac{\pi}{30} \text{ rad s}^{-1}.$$

The acceleration of a particle moving in a circle

Even when the magnitude of the velocity is constant, the direction of the velocity is constantly changing if the particle is moving in a circle. Since there is a change of velocity, there will be an acceleration, the measure of the rate of change of velocity.

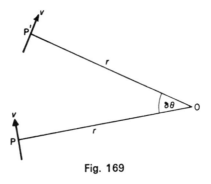

Fig. 169

Consider a particle when it has travelled a distance $r\,\delta\theta$ from an initial position P to a subsequent position P′ in a small interval of time δt. If OQ is the bisector of the angle POP′, the component of velocity perpendicular to OQ is $v \cos \frac{1}{2}\delta\theta$ when the particle is at P and when it is at P′, so there is no acceleration perpendicular to OQ. But whereas there was a component $v \sin \frac{1}{2}\delta\theta$ along $\overrightarrow{OQ}$ when the particle was at P, there is a component $-v \sin \frac{1}{2}\delta\theta$ along $\overrightarrow{OQ}$ when the particle is at P′, after δt seconds.

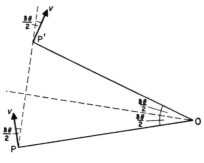

Fig. 170

In the direction $\overrightarrow{QO}$ then, there is a change of velocity of $2v \sin \frac{1}{2}\delta\theta$ in δt seconds. The acceleration, therefore, along $\overrightarrow{QO}$ is the limit of

$$\frac{2v \sin \frac{1}{2}\delta\theta}{\delta t}$$

as $\delta t \to 0$. But $\delta\theta = \omega\delta t$, where ω is the angular velocity. Hence

$$\text{Acceleration} = \frac{2v \sin \frac{1}{2}\delta\theta}{\dfrac{\delta\theta}{\omega}}$$

$$= v\omega \left(\frac{\sin \frac{1}{2}\delta\theta}{\frac{1}{2}\delta\theta} \right)$$

It can be proved that as $\delta\theta \to 0$,

$$\frac{\sin \delta\theta}{\delta\theta} \to 1$$

Hence $\qquad$ acceleration along $\overrightarrow{QO} = v\omega$

$$= \frac{v^2}{r} \text{ or } r\omega^2.$$

Thus a particle travelling in a circle radius r with constant speed v has an acceleration v^2/r inwards to the centre of the circle.

Application of Newton's Second Law

When a body describes a circle, we have seen that there is an acceleration towards the centre of the circle. Since there is an acceleration, there must be a force to produce that acceleration. In practice, if we think of various bodies describing circles, in many cases the nature of the force acting on the body to produce that acceleration is immediately apparent. A stone tied to a length of string can be whirled round in a circle: the tension in the string provides the necessary force. An aircraft describing a circle may bank so that extra force acts inwards to the centre to produce the acceleration required. A body placed on a gramophone

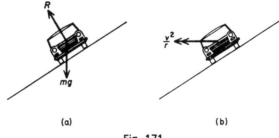

(a) $\qquad\qquad$ (b)

Fig. 171

turntable moves in a circle if the body and turntable are sufficiently rough to give a force large enough to produce the central acceleration. A road with a tight corner may be banked so that a component of the normal reaction can cause the acceleration, which then does not depend on friction.

Example 1. *A body mass 2 kg is describing a circle radius 0.5 m with velocity 4 m s^{-1}. Find the force causing the body to move in a circle.*

The acceleration of the body is given by

$$\frac{v^2}{r} = \frac{4^2}{0.5} = 32 \text{ m s}^{-2}.$$

Newton's Second Law gives

$$\text{Force} = 2 \times 32 = 64 \text{ N.}$$

The force is 64 N towards the centre of the circle.

Example 2. *A body mass 3 kg is describing a circle radius 0.3 m at 90 r.p.m. Find the force causing the circular motion.*

$$90 \text{ r.p.m.} = 90 \times \frac{2\pi}{60} \text{ rad s}^{-1}$$

$$= 3\pi \text{ rad s}^{-1}.$$

The acceleration of the body is

$$r\omega^2 = 0.3 \times (3\pi)^2 = 2.7\,\pi^2$$

Hence the force on the body is

$$3 \times (2.7\pi^2) = 80 \text{ N, to 2 significant figures.}$$

In these examples we have found the force which makes the body move in a circle. If there was no resultant force acting on the body, it would move in a straight line (Newton's First Law). There may be many forces acting on a body, but if it is to describe a circle radius r with constant speed v (i.e. constant angular velocity ω), the resultant of all the forces acting on the body must produce the necessary acceleration v^2/r (or $r\omega^2$) towards the centre of the circle. If the data of a problem are at all complicated, it is advisable to draw two diagrams, one for the forces acting on the body and the second for the acceleration which these forces combine to produce.

Example 3. *A boy mass 40 kg at a fairground is on a roundabout rotating at 1 rad s^{-1}. He is standing 5 m from the axis of the roundabout. Find the force causing him to move in a circle.*

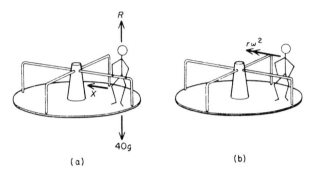

Fig. 172

The circle is in a horizontal plane and the only horizontal force acting on the boy is the frictional force X, so

$$F = ma$$
$$\Rightarrow X = 40(5 \times 1^2)$$
$$= 200.$$

The force causing circular motion is 200 N.

Since the boy has not a vertical acceleration,

$$R - 40g = 0$$
$$\Rightarrow R = 40 \times 9.8$$
$$= 392.$$

The normal reaction is 392 N.

Since the boy describes a circle and the force is provided by friction,

$$X \leqslant \mu R$$
$$\Rightarrow 200 \leqslant \mu. 392$$
$$\Rightarrow \mu \geqslant 200/392$$
i.e. $$\mu \geqslant 0.51$$

The coefficient of friction for this circular motion to be possible must be at least 0.51.

Conical pendulum

Example. *One end of a light cord 1.2 m long is fixed at a point P: to the other end is attached a stone of mass 0.2 kg. The stone is whirled in a horizontal circle at 5 rad s⁻¹. Find the tension in the string and the distance below P of the horizontal plane containing the circle.*

(If possible, set up this simple experiment to see the plane of the circle in relation to P.)

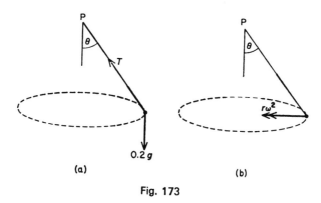

(a)   (b)

Fig. 173

Since the vertical acceleration of the body is zero

$$T \cos \theta = 0.2g$$

Since the horizontal acceleration is $1.2 \sin \theta \times 5^2$, and the horizontal component of the tension is $T \sin \theta$,

$$T \sin \theta = 0.2 \times 1.2 \sin \theta \times 5^2$$
$$\Rightarrow T = 0.2 \times 1.2 \times 5^2$$

i.e. $$T = 6.$$

Also $$T \cos \theta = 0.2g$$

i.e. $$6 \cos \theta = 0.2g,$$

$$\cos \theta = g/30$$

The distance of the plane of the circle below P is

$$1.2 \cos \theta = 1.2 \times \frac{g}{30}$$

$$= 0.392 \text{ m}.$$

The tension in the string is 6 N and the plane of the circle is 0.392 m below P. This system is called a *conical pendulum*.

EXERCISE 26

1. A body mass 1 kg is describing a circle radius 1 m with an angular velocity of 1 rad s^{-1}. Find the force towards the centre of the circle.

2. A body mass 4 kg is describing a circle radius 0.6 m at 50 revolutions per minute. Find the force towards the centre of the circle.

3. Find the central force necessary if a body mass 4.8 kg describes a circle radius 0.5 m at 5 rad s^{-1}.

4. A cyclist mass 70 kg describes a horizontal circle radius 20 m at 5 m s^{-1}. What is the frictional force which acts towards the centre of the circle?

5. A body of mass 5 kg is describing a circle radius 0.2 m. If the central force on the body is 16 N, find the angular velocity of the body, in radians per second.

6. A piece of metal of mass 0.1 kg is fixed to a spoke of a bicycle wheel, 12 cm from the axle. If the difference between the tensions on either side of the metal is 30 N, find the angular velocity of the wheel.

7. A flywheel with radius 10 cm is rotating at a constant speed of 3600 r.p.m. Find the acceleration of a point on the rim of the flywheel. (Take π^2 to be 10.)

8. An insect of mass 6 grammes is at a point X on the rim of a flywheel, radius 12 cm, rotating at 250 r.p.m. What force must the flywheel exert on the insect if it stays at X?

9. A bead of mass 20 grammes is projected with velocity 5 m s^{-1} along a smooth circular tube with radius 0.4 m, which is fixed in a horizontal plane. Calculate
   (i) the horizontal force on the bead,
   (ii) the vertical force exerted by the tube on the bead,
   (iii) the total force exerted by the tube on the bead.

10. One end of a light string of length 0.4 m is fixed at a point A. To the other end is attached a mass of 3 kg. The mass is made to describe a horizontal circle at 7 rad s^{-1}. Find the inclination of the string to the vertical, and the tension in the string.

11. One end of a light string of length 10 cm is fixed at a point A. To the other end is attached a stone, which is made to describe a horizontal circle. If the inclination of the string to the vertical is 60°, find the angular velocity of the stone.

12. One end of a light string is fixed at a point. To the other end is attached a body of mass 3 kg. If the tension in the string must not exceed 70 N, find the greatest angle which the string can make with the vertical.

13. One end of a light string of length 1 m is fixed at a point O. To the other end is attached a mass of 3 kg. If the tension in the string is 30 N, find the angular velocity with which the mass will be describing a horizontal circle in a plane through O.

14. A car of mass 700 kg is travelling along a road in the arc of a circle radius 80 m at 20 m s^{-1}. Find the component towards the centre of the force exerted by the road on the car.

15. A train is running along a track in the arc of a circle of radius 250 m at 17.5 m s^{-1}. Find the angle at which the rails must be banked if there is no sideways thrust on them.

16. A small metal disc of mass 0.02 kg is attached to one end of a light string, length 0.4 m. The other end of the string is fixed at a point P, and the disc is made to describe a horizontal circle centre P at 4 rad s^{-1}. The string encounters a peg Q placed 0.3 m from P, and begins to describe a circle, centre Q. Assuming that there is no loss of momentum when the string hits the peg, calculate the change in the tension in the string at that moment.

17. Two equal particles are connected by a string which passes through a hole in a smooth table. One particle lies on the table and the other hangs vertically. What angular velocity must be given to the particle on the table to keep it moving in a circle of radius 20 cm, while the other particle remains at rest?

18. A stone of mass 0.1 kg describes a horizontal circle on a horizontal table at the end of a string of length 0.8 m, the other end of the string being fixed at a point on the table. The angular velocity of the stone is 4 rad s^{-1}. It collides with a lump of plasticine, mass 0.06 kg, which adheres to the stone. Calculate

(i) the velocity with which the stone and the plasticine begin to move,

(ii) the loss of kinetic energy,

(iii) the change in the tension in the string.

19. In a certain machine at a funfair, a man stands against the wall of a circular room, radius 3 m. The room rotates about a vertical axis through the centre of the floor at 2 rad s^{-1}, then the floor is

lowered. If the man does not slip vertically downwards, find the least value of the coefficient of friction between the man and the wall.

20. A satellite is circling the Earth at a height of 800 km. Find the speed of the satellite. (Take the Earth to be a sphere of radius 6400 km, and neglect variation in gravity.)

CHAPTER 17

Further Vectors

Units

The mass of a box of chocolates may be 200 grams; the mass is described completely by a number and the unit of measurement. Similarly with other scalar quantities such as time, temperature, and kinetic energy. Vectors, we recall, need the direction specified as well as the magnitude, or else must be determined by the components along each axis. If we are using unit vectors **i** and **j** a force can be described completely as, say $(3\mathbf{i} + 4\mathbf{j})$ N, but if we are using matrices to represent vectors, then writing the units each time beside the matrix is rather confusing, so that we usually state at the beginning that the units are SI, that force is measured in newtons, lengths in metres, etc; we write that a force $\mathbf{F} = \begin{pmatrix} 3 \\ 4 \end{pmatrix}$ and we do not usually write $\mathbf{F} = \begin{pmatrix} 3 \\ 4 \end{pmatrix}$ N nor that a velocity $\mathbf{v} = \begin{pmatrix} 3 \\ 4 \end{pmatrix}$ m s^{-1}.

Example 1. *The position vector at time t seconds of a body mass 5 kg is $10t^2\mathbf{i} + 4\mathbf{j}$, the unit of distance being the metre. Find the velocity of the body after 2 seconds and the force acting on the body.*

If **r** is the position vector of the body,
$$\mathbf{r} = 10t^2\mathbf{i} + 4\mathbf{j}$$
If **v** is the velocity vector, $\mathbf{v} = d\mathbf{r}/dt$,
$$\mathbf{v} = 20t\mathbf{i}$$

thus the velocity after 2 seconds is 40 m s^{-1} parallel to the x-axis.

Since
$$\mathbf{F} = m\mathbf{a} \quad \text{and} \quad \mathbf{a} = d\mathbf{v}/dt,$$
$$\mathbf{F} = 5(20\mathbf{i}),$$

so that the force is constant, 100 N parallel to the x-axis.

Notice that we could have used matrices to describe these vectors, writing

$$\mathbf{r} = \begin{pmatrix} 10t^2 \\ 4 \end{pmatrix}, \quad \mathbf{v} = \begin{pmatrix} 20t \\ 0 \end{pmatrix}, \quad \text{and} \quad \mathbf{F} = \begin{pmatrix} 100 \\ 0 \end{pmatrix}$$

Example 2. *The position vectors of particles A, B at time t are $2t^3\mathbf{i} + 2t^2\mathbf{j}$, $t^3\mathbf{i} + 4t^2\mathbf{j}$ respectively. Find the velocity of A relative to B and the speed of A relative to B after 1 second.*

Since the position of A relative to B is described by the vector $t^3\mathbf{i} - 2t^2\mathbf{j}$, the velocity of A relative to B is $3t^2\mathbf{i} - 4t\mathbf{j}$. The speed of A relative to B is $\sqrt{(9t^4 + 16t^2)}$, so that the speed after 1 second is 5 m s^{-1}.

Throughout all the exercises in this chapter, the unit of force is the newton, of length is the metre and of time is the second.

EXERCISE 27

1. A body mass 2 kg, initial velocity $3\mathbf{i}$, is acted on by a force $\mathbf{F}$. Find the velocity and the speed after 2 seconds if
   (a) $\mathbf{F} = 4\mathbf{i}$,
   (b) $\mathbf{F} = -4\mathbf{i}$,
   (c) $\mathbf{F} = 4\mathbf{j}$.
2. A body mass 5 kg, initial velocity $\mathbf{i} + 2\mathbf{j}$, is acted on by a force $\mathbf{F}$. Find the velocity and speed of the body after 2 seconds if
   (a) $\mathbf{F} = 5\mathbf{j}$,
   (b) $\mathbf{F} = -5\mathbf{j}$,
   (c) $\mathbf{F} = 5\mathbf{i} + 5\mathbf{j}$.
3. The velocity $\mathbf{v}$ at time t of a body mass 3 kg is given by $\mathbf{v} = 4t\mathbf{i} + \mathbf{j}$. Find the force acting on the body.
4. The velocity $\mathbf{v}$ of a body mass 5 kg is given by $\mathbf{v} = 2t\mathbf{i} + t^2\mathbf{j}$. Find the force acting on the body.
5. A force $4\mathbf{i} + 3\mathbf{j}$ produced an acceleration $0.8\mathbf{i} + 0.6\mathbf{j}$ in a certain body. Find the mass of that body.
6. A force $500\mathbf{i} + 800\mathbf{j}$ acting on a certain body produces an acceleration $10\mathbf{i} + 16\mathbf{j}$ in that body. Find the mass of the body.
7. Show that, whatever the mass of the body, a force $4\mathbf{i} + 3\mathbf{j}$ acting on a body cannot produce an acceleration $0.6\mathbf{i} + 0.8\mathbf{j}$.

8. Two forces $\mathbf{F_1}$ and $\mathbf{F_2}$ acting on a body mass 5 kg produce an acceleration $0.2\mathbf{i} + 0.3\mathbf{j}$. Find $\mathbf{F_2}$ if $\mathbf{F_1} = 2\mathbf{i} + \mathbf{j}$.

9. A body mass 5 kg initial velocity $2\mathbf{i}$ is acted on by a force $\mathbf{F}$. Find the velocity at time $t = 2$ if
   (a) $\mathbf{F} = 30\mathbf{i} + 20\mathbf{j}$,
   (b) $\mathbf{F} = 20t\mathbf{i} + 10\mathbf{j}$,
   (c) $\mathbf{F} = 30t^2\mathbf{i} + 10t\mathbf{j}$.

10. A body mass 2 kg has position vector $\mathbf{r}$. Find the force $\mathbf{F}$ acting on the body if
    (a) $\mathbf{r} = t^2\mathbf{i} + t\mathbf{j}$,
    (b) $\mathbf{r} = (t - 1)^2\mathbf{i} + (t - 1)\mathbf{j}$,
    (c) $\mathbf{r} = (t + 1)^2\mathbf{i} + (t + 1)\mathbf{j}$.
    Comment on your results.

11. The position vector $\mathbf{r}$ of a particle P is $\begin{pmatrix} 5t^2 \\ t \end{pmatrix}$. Find in matrix form the velocity and acceleration of P.

12. The velocity of a particle P at time t is $\begin{pmatrix} 3t^2 \\ 4t \end{pmatrix}$. If initially the position vector of P is $\begin{pmatrix} 2 \\ 0 \end{pmatrix}$, find the position vector and acceleration vector of P.

13. The acceleration of a particle P is given by the vector $\begin{pmatrix} 0 \\ -10 \end{pmatrix}$. If initially P has position vector $\begin{pmatrix} 5 \\ 0 \end{pmatrix}$ and velocity vector $\begin{pmatrix} 40 \\ 50 \end{pmatrix}$, find in matrix form the velocity vector and the position vector of P at time t.

14. A body mass 5 kg velocity $3\mathbf{i} + 7\mathbf{j}$ receives an impulse $\mathbf{I}$ which changes its velocity to $\mathbf{V}$. Find the magnitude of $\mathbf{I}$ if
    (a) $\mathbf{V} = 3\mathbf{j}$,
    (b) $\mathbf{V} = 3\mathbf{i}$,
    (c) $\mathbf{V} = 3\mathbf{i} + 3\mathbf{j}$.

15. A body, mass 2 kg, velocity $2\mathbf{i} + 5\mathbf{j}$, receives an impulse $\mathbf{I}$ which changes its velocity to $\mathbf{V}$. Find $\mathbf{I}$ if
    (a) $\mathbf{V} = 2\mathbf{i} + 2\mathbf{j}$,
    (b) $\mathbf{V} = 2\mathbf{i}$,
    (c) $\mathbf{V} = -2\mathbf{i} + 2\mathbf{j}$.

16. At time t a particle P has position vector $t^2\mathbf{i} + t\mathbf{j}$ and a particle Q has position vector $2t\mathbf{i} - \mathbf{j}$.
    (a) Find the position vectors of P and Q when $t = 2$.
    (b) Show that Q has constant velocity, and find the velocity of P when $t = 2$.
    (c) Find the velocity of P relative to Q at time $t = 2$.

17. At time t a particle P has position vector $2t\mathbf{i} + (t + 1)\mathbf{j}$, and a particle Q has position vector $-3t\mathbf{i} + (2t - 3)\mathbf{j}$.
   (a) Find the position vectors of P and Q when $t = 1$.
   (b) Show that both P and Q have constant velocity.
   (c) Find the velocity of P relative to Q.

18. A particle P has position vector $\begin{pmatrix} 3 \\ 4 \end{pmatrix}$ and constant velocity vector $\begin{pmatrix} 1 \\ 2 \end{pmatrix}$ at the same time as a particle Q has position vector $\begin{pmatrix} -2 \\ 24 \end{pmatrix}$ and constant velocity vector $\begin{pmatrix} 2 \\ -2 \end{pmatrix}$. Find the position vector of each particle at time t, and show that the particles will collide.

19. A particle P has position vector $\begin{pmatrix} 1 \\ 2 \end{pmatrix}$ and constant velocity vector $\begin{pmatrix} 3 \\ 1 \end{pmatrix}$ at the same time as a particle Q has position vector $\begin{pmatrix} 10 \\ 10 \end{pmatrix}$ and constant velocity vector $\begin{pmatrix} 1 \\ -1 \end{pmatrix}$. Find whether the particles collide.

20. A cricket ball mass 0.15 kg reaches a batsman with velocity $10\mathbf{i}$ and is hit back with velocity of $-30\mathbf{i}$. Find the impulse given by the bat to the ball. If the bat and ball are in contact for 0.05 s, find the average force exerted by the bat on the ball.

Systems of forces; forces in equilibrium

A body is in equilibrium under the action of forces $\mathbf{F}_1$, $\mathbf{F}_2$, ... if the vector sum of the forces is zero,

i.e., $$\mathbf{F}_1 + \mathbf{F}_2 + \mathbf{F}_3 + \mathbf{F}_4 \cdots = \mathbf{0}*$$

and if the sum of the moments about any point is zero. This last condition will be met if the lines of action of the forces pass through a single point, as when the forces are acting on a small body or particle. Thus a small body can be in equilibrium under the action of forces $\begin{pmatrix} 3 \\ 4 \end{pmatrix}$, $\begin{pmatrix} 2 \\ 3 \end{pmatrix}$, $\begin{pmatrix} -5 \\ -7 \end{pmatrix}$ but not under forces $\begin{pmatrix} 3 \\ 4 \end{pmatrix}$, $\begin{pmatrix} 2 \\ 3 \end{pmatrix}$, $\begin{pmatrix} -5 \\ -6 \end{pmatrix}$ nor under forces $\begin{pmatrix} 3 \\ 4 \end{pmatrix}$, $\begin{pmatrix} 2 \\ 3 \end{pmatrix}$, $\begin{pmatrix} -4 \\ -7 \end{pmatrix}$.

* Abbreviated $\Sigma \mathbf{F} = 0$.

Conversely, if we know that a body is equilibrium under the action of forces $\begin{pmatrix} 3 \\ 4 \end{pmatrix}$, $\begin{pmatrix} x \\ -5 \end{pmatrix}$, $\begin{pmatrix} -2 \\ y \end{pmatrix}$

then
$$\begin{pmatrix} 3 \\ 4 \end{pmatrix} + \begin{pmatrix} x \\ -5 \end{pmatrix} + \begin{pmatrix} -2 \\ y \end{pmatrix} = \begin{pmatrix} 0 \\ 0 \end{pmatrix}$$

i.e.,
$$\begin{pmatrix} 3 + x - 2 \\ 4 - 5 + y \end{pmatrix} = \begin{pmatrix} 0 \\ 0 \end{pmatrix},$$
$$x = -1 \quad \text{and} \quad y = 1.$$

Resultant of a system of forces

If a single force **F** is the resultant of two forces $\mathbf{F}_1$ and $\mathbf{F}_2$, then $\mathbf{F} = \mathbf{F}_1 + \mathbf{F}_2$ and the line of action of **F** must pass through the intersection of the lines of action of $\mathbf{F}_1$ and $\mathbf{F}_2$.

Thus if points A, B, and C have position vectors $3\mathbf{i} + \mathbf{j}$, $2\mathbf{i} + 4\mathbf{j}$ and $\mathbf{i} + \mathbf{j}$ respectively (Fig. 174), and if forces $\mathbf{F}_1$ and $\mathbf{F}_2$ are represented completely by $\overrightarrow{CA}$ and $\overrightarrow{CB}$, then $\mathbf{F}_1 = 2\mathbf{i}$ and $\mathbf{F}_2 = \mathbf{i} + 3\mathbf{j}$ and the resultant is $3\mathbf{i} + 3\mathbf{j}$ through C.

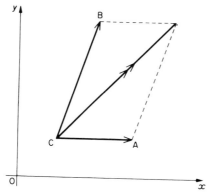

Fig. 174

Example. *Forces* $\mathbf{F}_1 = 3\mathbf{i} - 4\mathbf{j}$, $\mathbf{F}_2 = 2\mathbf{i} + 5\mathbf{j}$ *and* $\mathbf{F}_3 = x\mathbf{i} + y\mathbf{j}$
act through a point. Find x and y if

   (*i*) *the system is in equilibrium;*
   (*ii*) *the system is equivalent to a single force* 5$\mathbf{j}$;
   (*iii*) $\mathbf{F}_3$ *is the resultant of* $\mathbf{F}_1$ *and* $\mathbf{F}_2$.

Firstly, if $\mathbf{F}_1 + \mathbf{F}_2 + \mathbf{F}_3 = 0$,
$$(3\mathbf{i} - 4\mathbf{j}) + (2\mathbf{i} + 5\mathbf{j}) + (x\mathbf{i} + y\mathbf{j}) = 0,$$

i.e., $x = -5$ and $y = -1$.

Secondly, if $\mathbf{F}_1 + \mathbf{F}_2 + \mathbf{F}_3 = 5\mathbf{j}$,
$$(3\mathbf{i} - 4\mathbf{j}) + (2\mathbf{i} + 5\mathbf{j}) + (x\mathbf{i} + y\mathbf{j}) = 5\mathbf{j},$$

i.e., $x = -5$ and $y = 4$.

Thirdly, if $\mathbf{F}_1 + \mathbf{F}_2 = \mathbf{F}_3$,
$$(3\mathbf{i} - 4\mathbf{j}) + (2\mathbf{i} + 5\mathbf{j}) = (x\mathbf{i} + y\mathbf{j}),$$

i.e., $x = 5$ and $y = 1$.

EXERCISE 28

1. Are the following systems of forces, acting through a single point,
   in equilibrium?
   (a) $2\mathbf{i} + 3\mathbf{j}$, $3\mathbf{i} + 3\mathbf{j}$, $-5\mathbf{i} - 7\mathbf{j}$.
   (b) $3\mathbf{i} - 4\mathbf{j}$, $4\mathbf{i} + 4\mathbf{j}$, $-7\mathbf{i}$.
   (c) $2\mathbf{i}$, $\mathbf{i} + 3\mathbf{j}$, $-\mathbf{i}$, $-5\mathbf{j}$.
   (d) $3\mathbf{i} - 5\mathbf{j}$, $-4\mathbf{i} + 6\mathbf{j}$, $\mathbf{i} - \mathbf{j}$.
2. Find x and y if the following forces, acting through a single point,
   are in equilibrium:
   (a) $3\mathbf{i} + 4\mathbf{j}$, $-7\mathbf{i} - \mathbf{j}$, $x\mathbf{i} + y\mathbf{j}$.
   (b) $2\mathbf{i} - 5\mathbf{j}$, $x\mathbf{i} + 3\mathbf{j}$, $-6\mathbf{i} + y\mathbf{j}$.
   (c) $x\mathbf{i}$, $4\mathbf{j}$, $2\mathbf{i} - 6\mathbf{j}$, $y\mathbf{j}$.
   (d) $2\mathbf{i} - 5\mathbf{j}$, $-8\mathbf{i} - 3\mathbf{j}$, $2x\mathbf{i} + 2y\mathbf{j}$.
3. The magnitude of a force $\mathbf{F}$ is 20 newtons. Find $\mathbf{F}$ when
   (a) $\mathbf{F}$ is parallel to the unit vector $\mathbf{i}$,
   (b) $\mathbf{F}$ is parallel to the vector $3\mathbf{i} + 4\mathbf{j}$,
   (c) $\mathbf{F}$ is parallel to the vector $-4\mathbf{i} + 3\mathbf{j}$,
   (d) $\mathbf{F}$ is parallel to the vector $\mathbf{i} + \mathbf{j}$.
   Find also the two possible forces $\mathbf{F}$ if $\mathbf{F}$ is perpendicular to the
   vector $4\mathbf{i} - 3\mathbf{j}$.

4. The position vectors of points P and Q relative to an origin O are **2i** and **3i + 4j** respectively. Calculate the resultant of forces represented completely by OP and OQ, and find the third force Z which must be added to the system if it is to be in equilibrium. Through which point must Z act?

5. Two forces $\mathbf{F}_1$ and $\mathbf{F}_2$ are such that $\mathbf{F}_1$ is perpendicular to $\mathbf{F}_1 + \mathbf{F}_2$ and $\mathbf{F}_1$ is equal in magnitude to $\mathbf{F}_1 + \mathbf{F}_2$. Draw a diagram showing $\mathbf{F}_1$, $\mathbf{F}_2$ and $\mathbf{F}_1 + \mathbf{F}_2$, and hence find the ratio of the magnitudes of $\mathbf{F}_1$ and $\mathbf{F}_2$.

Vectors in three dimensions

It is difficult to describe direction in three dimensions, so that we usually use the components along mutually perpendicular axes Ox, Oy, and Oz and, if we need any angles, the angles made by the vector with those axes.

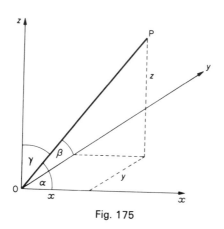

Fig. 175

Unit vectors **i, j, k** along the axes Ox, Oy, Oz respectively can be represented by matrices so that

$$\mathbf{i} = \begin{pmatrix} 1 \\ 0 \\ 0 \end{pmatrix} \qquad \mathbf{j} = \begin{pmatrix} 0 \\ 1 \\ 0 \end{pmatrix} \qquad \mathbf{k} = \begin{pmatrix} 0 \\ 0 \\ 1 \end{pmatrix}$$

The position vector of a point P, coordinates (x, y, z), can be written $x\mathbf{i} + y\mathbf{j} + z\mathbf{k}$ or represented by the matrix $\begin{pmatrix} x \\ y \\ z \end{pmatrix}$. From Fig. 175 we see that the length of OP is $\sqrt{(x^2 + y^2 + z^2)}$ and we call $\sqrt{(x^2 + y^2 + z^2)}$ the *magnitude* (or *modulus*) of the three-dimensional vector $x\mathbf{i} + y\mathbf{j} + z\mathbf{k}$. The angles α, β, and γ between OP and $\mathbf{i}$, $\mathbf{j}$, and $\mathbf{k}$ are such that, by trigonometry,

$$\cos \alpha = \frac{x}{\text{OP}}, \quad \cos \beta = \frac{y}{\text{OP}}, \quad \text{and} \quad \cos \gamma = \frac{z}{\text{OP}}.$$

Example 1. *Points A and B have position vectors relative to an origin O of* $\begin{pmatrix} 3 \\ 4 \\ 5 \end{pmatrix}$ *and* $\begin{pmatrix} 4 \\ 6 \\ 3 \end{pmatrix}$ *respectively. Find the length of AB and the angle between AB and the x-axis.*

Since $\overrightarrow{\text{OA}} = \begin{pmatrix} 3 \\ 4 \\ 5 \end{pmatrix}$ and $\overrightarrow{\text{OB}} = \begin{pmatrix} 4 \\ 6 \\ 3 \end{pmatrix}$, $\overrightarrow{\text{AB}} = \begin{pmatrix} 1 \\ 2 \\ -2 \end{pmatrix}$, the length of AB $= \sqrt{\{1^2 + 2^2 + (-2)^2\}} = 3$, and the angle α between AB and the x-axis is given by $\cos \alpha = \frac{1}{3}$, i.e., $\alpha = \text{arc cos} (\frac{1}{3})$, about $70.5°$.

Example 2. *The position vector at time t of a point P relative to an origin O is* $(\frac{1}{3}t^3 - t)\mathbf{i} + \mathbf{j} + t^2\mathbf{k}$. *Find the speed of the particle at time t.*

If $\mathbf{r}$ is the position vector and $\mathbf{v}$ the velocity vector,

$$\mathbf{r} = (\tfrac{1}{3}t^3 - t)\mathbf{i} + \mathbf{j} + t^2\mathbf{k}$$

and

$$\mathbf{v} = d\mathbf{r}/dt = (t^2 - 1)\mathbf{i} + 2t\mathbf{k}.$$

The speed is the magnitude of the velocity, so that

$$\begin{aligned} v &= \sqrt{\{(t^2 - 1)^2 + (2t)^2\}} \\ &= \sqrt{\{t^4 - 2t^2 + 1 + 4t^2\}} \\ &= t^2 + 1. \end{aligned}$$

The great advantage of working in vectors is that they enable problems in three dimensions to be solved as easily as problems in two dimensions.

Example 3. *Find the position vector of the centre of mass of particles mass m, 2m at points position vectors* $2\mathbf{i} + \mathbf{j} - 2\mathbf{k}$, $5\mathbf{i} - 2\mathbf{j} + 4\mathbf{k}$.

Proceeding as on page 129, if the position vector of the centre of mass is $\bar{\mathbf{r}}$,

$$(3m)\bar{\mathbf{r}} = m(2\mathbf{i} + \mathbf{j} - 2\mathbf{k}) + 2m(5\mathbf{i} - 2\mathbf{j} + 4\mathbf{k})$$
$$\therefore \quad \bar{\mathbf{r}} = 4\mathbf{i} - \mathbf{j} + 2\mathbf{k}$$

EXERCISE 29

1. Find the magnitude of each of the following vectors:

(a) $\begin{pmatrix} 3 \\ -4 \\ 5 \end{pmatrix}$, (b) $\begin{pmatrix} 2 \\ -1 \\ 2 \end{pmatrix}$, (c) $\begin{pmatrix} 2 \\ 6 \\ 9 \end{pmatrix}$, (d) $\begin{pmatrix} 12 \\ -4 \\ -3 \end{pmatrix}$.

2. If $\overrightarrow{OA} = \begin{pmatrix} 2 \\ 3 \\ 4 \end{pmatrix}$, $\overrightarrow{OB} = \begin{pmatrix} 0 \\ 2 \\ 2 \end{pmatrix}$, $\overrightarrow{OC} = \begin{pmatrix} 2 \\ -4 \\ -7 \end{pmatrix}$.

find the magnitudes of $\overrightarrow{AB}$, $\overrightarrow{BC}$, and $\overrightarrow{AC}$ and the cosine of the angle each vector makes with the *x*-axis.

3. If $\overrightarrow{OA} = \begin{pmatrix} 1 \\ 2 \\ 2 \end{pmatrix}$ and $\overrightarrow{OB} = \begin{pmatrix} 2 \\ 6 \\ -9 \end{pmatrix}$ find the cosine of the angles made

by $\overrightarrow{OA}$, $\overrightarrow{OB}$ and $\overrightarrow{AB}$ with the *x*-axis and the *y*-axis.

4. If the cosine of the angle between the vector $\overrightarrow{OP}$ and the *x*-axis is a and the cosine of the angle between the vector $\overrightarrow{OP}$ and the *y*-axis is b, find the cosine of the angle between $\overrightarrow{OP}$ and the *z*-axis.

5. A force $\mathbf{F} = 2\mathbf{i} + 6\mathbf{j} + 4\mathbf{k}$ acts on a body mass 2 kg. Find the acceleration of the body.

6. A body mass 5 kg, initial velocity $3\mathbf{i}$, is acted on by a force $10\mathbf{i} + 20\mathbf{j} + 15\mathbf{k}$. Find the velocity of the body after 5 seconds.

7. The velocity $\mathbf{v}$ at time t of a body mass 3 kg is given by $\mathbf{v} = 6t\mathbf{i} + \mathbf{j} + 2t^2\mathbf{k}$. Find the force acting on the body.

8. A body mass 5 kg has position vector, at time t, $\frac{1}{2}t^2\mathbf{i} + 2t^3\mathbf{j} + 3t^4\mathbf{k}$. Find the force acting on the body, and the speed of the body when $t = 0.5$.

9. A body mass 2 kg, velocity $2\mathbf{i} - 3\mathbf{j} - \mathbf{k}$, receives an impulse $8\mathbf{i} + 6\mathbf{j} + 4\mathbf{k}$. Find the subsequent velocity of the body.

10. A cricket ball mass 0.15 kg reaches a batsman with velocity $10\mathbf{i} - \mathbf{k}$, and receives an impulse $-1.2\mathbf{i} + 3\mathbf{j} + 0.12\mathbf{k}$. Find the velocity with which the ball leaves the bat.

    Compare the path of the ball after this impulse with the paths after an impulse of $-1.2\mathbf{i} + 0.15\mathbf{k}$ and after an impulse $-1.2\mathbf{i} + 6\mathbf{k}$, $\mathbf{i}$ being a unit vector along the line between the centres of the wickets, $\mathbf{j}$ a unit horizontal vector at right angles to $\mathbf{i}$ and $\mathbf{k}$ a unit vector vertically upwards.

11. Forces $\mathbf{F}_1$, $\mathbf{F}_2$, and $\mathbf{F}_3$ are described by the vectors $2\mathbf{i} + 3\mathbf{j} - \mathbf{k}$, $2\mathbf{i} - 4\mathbf{j} + 3\mathbf{k}$, and $\mathbf{i} + 3\mathbf{k}$ respectively. Find the resultant of the following systems of forces:
    (a) $\mathbf{F}_1 + \mathbf{F}_2 + \mathbf{F}_3$,
    (b) $\mathbf{F}_1 + 2\mathbf{F}_2 + \mathbf{F}_3$,
    (c) $\mathbf{F}_1 - \mathbf{F}_2 - \mathbf{F}_3$,
    (d) $\mathbf{F}_1 - 2\mathbf{F}_2 + 3\mathbf{F}_3$.

12. If $\mathbf{F}_1 = 3\mathbf{i} + 2\mathbf{j} + \mathbf{k}$, $\mathbf{F}_2 = -2\mathbf{i} + \mathbf{k}$, and $\mathbf{F}_3 = 8\mathbf{i} + 4\mathbf{j} + \mathbf{k}$, find the magnitude of the following:
    (a) $\mathbf{F}_1 + \mathbf{F}_2$,
    (b) $\mathbf{F}_1 + \mathbf{F}_2 + \mathbf{F}_3$,
    (c) $\mathbf{F}_1 + 2\mathbf{F}_2 - \mathbf{F}_3$,
    (d) $\mathbf{F}_1 - \mathbf{F}_2 + \mathbf{F}_3$.

13. Particles P and Q have velocities $2\mathbf{i} + 3\mathbf{j} - 5\mathbf{k}$ and $2\mathbf{i} - 3\mathbf{j} + 5\mathbf{k}$ respectively. Find the velocity of P relative to Q and the velocity of Q relative to P.

14. A particle P has position vector $3\mathbf{j} - \mathbf{k}$, velocity vector $\mathbf{i} + \mathbf{j} + \mathbf{k}$, at the same time as particle Q has position vector $-3\mathbf{i} - 3\mathbf{j} + 5\mathbf{k}$ and constant velocity vector $2\mathbf{i} + 3\mathbf{j} - \mathbf{k}$. Find whether the particles collide, and illustrate the paths of the particles in a diagram.

15. Find the centre of mass of each of the following systems of particles:
    (a) mass m, position vector $\mathbf{i} + \mathbf{j} - \mathbf{k}$,
        mass $2m$, position vector $4\mathbf{i} + \mathbf{j} + 2\mathbf{k}$.
    (b) mass m, position vector $2\mathbf{j} - \mathbf{k}$,
        mass $3m$, position vector $4\mathbf{i} - 2\mathbf{j} + 3\mathbf{k}$.
    (c) mass m, position vector $2\mathbf{i} + \mathbf{k}$,
        mass $2m$, position vector $2\mathbf{j} - 3\mathbf{k}$,
        mass m, position vector $6\mathbf{i} + 5\mathbf{k}$.

Miscellaneous Exercise 1

1. The mechanism of a clockwork toy lorry can be assumed to exert a constant propulsive force under all conditions. Starting from rest on a level table, the lorry will attain a speed of 1.2 m s^{-1} in 3 seconds. The mass of the lorry is 0.15 kg.
   (a) Find the propulsive force in newtons.
   (b) If the lorry is loaded with blocks of total mass 0.3 kg, find how long it will take to reach a speed of 1.2 m s^{-1} from rest.
   (c) If the unloaded lorry is placed on a plank sloping at an angle whose sine is $\frac{1}{49}$, find how long it will take to reach the same speed, travelling up the plank.
   (d) Show that the loaded lorry cannot climb the plank.
                                                        (O. & C., S.M.P.)
2. The velocity v of a particle is given in metres per second by
   $$v = 10 - 7t + t^2$$
   where t is the time in seconds after the motion begins.
   (a) What is the initial velocity of the particle?
   (b) What is the velocity when $t = 10$?
   (c) When is the particle travelling towards the point from which it started?
   (d) How far does it travel in the first second?
   (e) What is its acceleration when $t = 7$?
   (f) When is the particle momentarily at rest?
3. A toy motor-car of mass 0.4 kg is propelled from rest by a clockwork engine which provides a tractive force of 2.7×10^{-3} N lasting for 8 s. Find
   (a) the acceleration produced,
   (b) the speed of the car at the end of this time.
   If the speed then decreases uniformly to zero during the next 32 s, find how far the car travels from start to finish   (M.E.I.)
4. A car A, moving along a straight road with a uniform acceleration of a m s^{-2}, passes a point O with velocity V m s^{-1}. State its velocity and position n seconds after passing O. Find also its velocity when it is $5V^2/8a$ metres from O.
   At the same time as car A passes O a second car B, travelling in the same direction with a constant velocity of $5V$ metres per second also passes O. Find in terms of V and a the distance from O of the point at which A overtakes B.   (J.M.B.)
5. A stone is thrown down a pitshaft 500 m deep. If the initial velocity of the stone is 1 m s^{-1}, find
   (a) the time taken to reach the bottom of the shaft,
   (b) the speed of the stone when it reaches the bottom.

6. A racing car, moving with constant acceleration, covers two successive kilometres in 30 s and 20 s respectively. Find the acceleration of the car, and the average speed for each kilometre.

7. A monorail car is to take passengers from airport to city centre, a distance of 12 km, and is designed to have a maximum speed of 120 km h^{-1}. Acceleration and deceleration are both uniform and the numerical value of the deceleration is double that of the acceleration. The total time for the journey is $7\frac{1}{2}$ minutes. Draw a velocity-time diagram to illustrate the motion of the car.
Hence, or otherwise, find
(a) for what time, in minutes, the car travels at maximum speed,
(b) the acceleration in m s^{-2}.   (M.E.I.)

8. The distance between two underground railway stations X and Y is 900 m. A train starts from X and accelerates uniformly until it reaches its maximum speed of 20 m s^{-1}. It maintains this constant speed for a time and then decelerates uniformly to stop at Y.
The ratios of the times taken during the periods of acceleration, maximum speed and deceleration are 2 : 3 : 1. Find the total time taken for the journey from X to Y.   (C*)

9. An electric train travels from station A to station B. The train accelerates uniformly from rest at A to a maximum speed of 15 m s^{-1}, then travels at this speed for 1 minute before slowing down uniformly to rest at B. The acceleration is half the retardation and the time for the whole journey is 2 minutes. Calculate
(a) the acceleration,
(b) the distance between the stations A and B.   (A.E.B.*)

10. The speed of a train between stations is given by the following table:

Time (minutes)	1	2	3	4	5	6	7
Speed (m s^{-1})	10	20	30	30	30	15	0

Draw a speed-time graph and from it find the acceleration of the train and the distance between the stations.

11. In the rectangle ABCD it is given that AB = 3 cm and BC = 2 cm. Forces of 6 N, 2 N, 3 N and 4 N act along the sides $\overrightarrow{AB}$, $\overrightarrow{BC}$, $\overrightarrow{DC}$ and $\overrightarrow{AD}$. Calculate the sums of the moments about each of the points A, B, C and D. What do you deduce about the resultant of these four forces?
Calculate the magnitude of this resultant.   (C*.)

12. (a) Forces of 13 N, 10 N, and 13 N act along the sides of a triangle AB, BC, and CA in the directions indicated by the order of the

letters. If AB = AC = 6.5 m and BC = 5 m, prove that the system of forces is equivalent to a couple and find the magnitude of the couple.

(b) ABC is an isosceles triangle, right-angled at A. Forces of 4 N, 3 N and 2 N respectively act along the sides AB, BC and CA of the triangle. Find the magnitude of the resultant of these three forces and the angle which the resultant makes with the side AB.

(A.E.B.*)

13. In a rectangle ABCD the side AB is of length 8 cm and AD of length 6 cm. Forces 1 N, 2 N, 3 N, 4 N, 5 N and 5 N act along AB, BC, CD, DA, AC and BD respectively, in the directions indicated by the letters. Calculate the magnitude and the direction of the resultant and the point where its line of action cuts AB produced.

(A.E.B.*)

14. ABC is a triangle with AB = 3 cm, BC = 6 cm, angle ABC = 45°; a point D, on CA produced, is such that CA = AD. Forces act along the directions of BC, BD and DC equal in newtons to the lengths of the lines in cm. Find by drawing the magnitude of the resultant of the two forces acting along BC and BD, and also of the resultant of the two forces acting along BD and DC.

(O & C.*)

15. The figure ABCDEF is a regular hexagon. Forces of 3 N, 2 N and 1 N act along the sides AB, BC, CD respectively. Find, preferably by drawing, the magnitude of the resultant of the three forces.

(O. & C.*)

16. Calculate the magnitude of the resultant of forces of 5 N and 7 N acting at an angle of 120°. Calculate also the angle between the resultant and the force of 7 N.   (O. & C.*)

17. A bucket holding 10 kg of water is attached to the rope in a well. The mass of the bucket is 5 kg. Find the tension in the rope when the bucket full of water
   (a) is at rest,
   (b) is rising with an acceleration of 0.4 m s^{-2},
   (c) is rising at a constant speed of 1 m s^{-1},
   (d) is rising but with a retardation of 0.4 m s^{-2}.

18. A water-skier can only perform effectively if the thrust of the water acts along the line of her body, assumed straight. If her mass is 60 kg, find the tension in the horizontal towing rope if she leans back at 10° to the vertical in steady motion.

What will be her acceleration (assumed horizontal) if the tension is doubled, but she leans back at 15°?

(Solution by drawing is recommended.)   (S.M.P.)

19. A man of mass 100 kg is hanging by a rope attached to a winch in a helicopter which is hovering. The helicopter begins to ascend vertically with an acceleration of 0.75 m s^{-2}. Determine the tension in the rope in newtons.

If the rope were to be wound in at a constant rate of 1 m s^{-1} what change would there be in the value of the tension in the rope? Comment briefly.

After a short time the helicopter ceases to rise vertically but subsequently moves horizontally with an acceleration of 0.75 m s^{-2}. It is observed that with the winch not operating, the rope hangs at a constant angle θ to the vertical where $\tan \theta = \frac{3}{4}$. What is the tension in the rope now, and what is the horizontal force on the man due to the resistance of the air?   (M.E.I.)

20. A car travelling at 90 km h^{-1} is brought to rest in half a kilometre. Calculate, in newtons per tonne, the force exerted by the brakes and find the time during which the brakes act.   (A.E.B.*)

21. A lift is raised with an initial acceleration of 2 m s^{-2} until its velocity is 6 m s^{-1}. The velocity then remains constant until the lift is brought to rest at the top with a retardation of 3 m s^{-2}. Calculate in newtons, the thrust between the floor of the lift and the feet of an 80 kg passenger during each stage of the ascent.

If the height of the shaft is 45 m, draw a velocity-time diagram for the motion and hence or otherwise find the total time taken for the ascent.   (M.E.I.)

22. A body mass 2 kg is acted on by forces $(2\mathbf{i} + 3\mathbf{j})$ N, $(4\mathbf{i} + 7\mathbf{j})$ N and $(\mathbf{i} + \mathbf{j})$ N. Find
   (a) the resultant force on the body,
   (b) the resultant acceleration of the body,
   (c) how far the body has moved during 3 s, if it was initially at rest.

23. A force $4/(t + 1)^3$ N acts on a body mass 1 kg. If the body was initially at rest, use $a = \mathrm{d}v/\mathrm{d}t$ to find the velocity of the body after 4 s, and the distance travelled in that time.

24. A force $1/(2t + 1)^3$ N acts on a body mass 4 kg. If the body initially had a velocity of 3 m s^{-1}, find the velocity of the body after 5 s, and the distance travelled in that time.

25. A cricket ball mass 0.14 kg moving with velocity $(30\mathbf{i})$ m s^{-1} is in contact with the bat for 0.1 s. If the subsequent velocity of the ball is $(20\mathbf{i} + 40\mathbf{j})$ m s^{-1}, find the magnitude of the force exerted by the bat on the ball.

26. A ladder leaning against a wall is about to slip. The coefficient of friction between the top of the ladder and the wall is the same as the coefficient of friction between the foot of the ladder and the floor. Show that the lines of action of the resultant reactions at the top and foot of the ladder are at right angles. If the ladder is uniform, show that the angle the ladder makes with the vertical is twice the angle of friction.

27. A body of mass 10 kg is placed on a rough plane inclined at 45° to the horizontal. If the coefficient of friction is $\frac{1}{2}$, find the least force necessary to move the body up the plane
    (a) when the force acts horizontally,
    (b) when the force acts parallel to the plane,
    (c) if the force can act in any direction.

28. A mass of 12 kg just slips on a rough inclined plane when the inclination of the plane is 30°. If the inclination is reduced to 20°, calculate the force acting along the line of greatest slope which will
    (a) just move the mass up the plane,
    (b) just move the mass down the plane.
    If the force obtained in (a) acts on the mass down the plane along the line of greatest slope, determine the acceleration of the mass and the distance moved by the mass from rest in the first $\frac{1}{2}$ s.
    (A.E.B.*)

29. A mass of 4 kg on a rough horizontal table is connected by a string passing over a smooth pulley at the edge of the table to another mass of 10 kg hanging freely. The masses move with an acceleration of 6.3 m s⁻². Find the coefficient of friction between the 4 kg mass and the table.

30. A 10 kg package starts to slide down a chute at 1 m s⁻¹. The inclined portion of the chute is 5 m in length and makes an angle of 30° with the horizontal. The join of the inclined and the horizontal portions of the chute is curved so that there is no check of the package at this point. If the coefficient of friction between the package and the chute is 0.2 along the whole length of the chute, find how far the parcel travels along the horizontal portion of the chute before coming to rest. (M.E.I.*)

31. A block of mass 10 kg rests on a rough horizontal table, the coefficient of friction being 0.2. A horizontal string attached to the block passes over a smooth pulley at the edge of the table and a light scale-pan hangs on the end of the string. Equal masses of m kg are now added to the block and to the scalepan and the block just begins to move. Find the numerical value of m.
    If instead of m kg, masses of 5 kg were used, what would now be the acceleration of the block? (C.*)

32. Find the momentum gained by a stone mass 2 kg after it has fallen for 5 s. If the momentum gained is only 95 N s, find the average air-resistance to the motion of the stone.

33. A car mass 750 kg accelerated from 10 m s^{-1} to 25 m s^{-1} in 7.5 s. What was the average accelerating force?

34. A truck of mass 20 000 kg moving with a velocity of 5 m s^{-1} strikes a stationary truck of mass 30 000 kg. If the trucks move on together, find their common velocity.

Determine the loss of kinetic energy due to the collision.

(A.E.B.)

35. Two bodies mass 2 kg and 1 kg are moving in the same straight line. The body mass 2 kg, moving at 15 m s^{-1} overtakes and adheres to the mass of 1 kg. They move on together with a velocity of 12 m s^{-1}. Calculate the initial velocity of the 1 kg mass and the loss of kinetic energy in the collision.   (A.E.B.*)

36. A body mass 1 kg is projected upwards with a velocity of 7.7 m s^{-1}. Find
   (a) the time before it is first at a height of 2.8 m,
   (b) the change in the momentum of the body between projection and this time.

37. A car of mass 1200 kg accelerates on a level road from 8 m s^{-1} to 30 m s^{-1} in 11 s. If the road resistance is 800 N, calculate the tractive force of the engine, assumed constant.   (O. & C.*)

38. A gun of mass 1400 kg, free to recoil, fires a shell mass 8 kg horizontally, with an actual velocity of 280 m s^{-1}. Find
   (a) the initial velocity of the gun,
   (b) the total kinetic energy produced by the explosion,
   (c) the work done by the recoil of the gun.   (O. & C.*)

39. A boy of mass 50 kg dives from the stern of a rowing boat of mass 100 kg. The boat is motionless in the water but free to move. If the boy gives himself a horizontal velocity of 3 m s^{-1} relative to the boat, find the horizontal velocity given to the boat. Find also the kinetic energy generated, assuming that neither the boat nor the boy has any vertical motion.   (C.*)

40. A pile-driver of mass 500 kg is dropped from a height of 10 m onto a pile of mass 200 kg and the pile sinks 0.1 m. Assuming that the pile-driver and the pile move on together after the impact, find
   (a) their velocity immediately after the impact,
   (b) the decrease in total energy of the pile and pile-driver in coming to rest after impact,
   (c) the average resistance in newtons.   (C.*)

41. A jet of water is directed vertically upwards at a speed of 3.5 m s^{-1} through a nozzle of cross-section 6×10^{-4} m^2. A ball is balanced in the air 0.6 m above the level of the nozzle by the impact of the

water on its underside. Assuming that all the water is momentarily brought to rest by the impact, calculate
(a) the velocity of the water on impact,
(b) the loss of momentum of the water per second,
(c) the mass of the ball.
(The mass of 1 m³ of water is 10³ kg.)   (C.*)

42. A boat mass 250 kg is moving ahead under its own momentum at a speed of 2 m s⁻¹ when a man mass 100 kg starts to run forward from stern to bow, running at a steady speed of 3·5 m s⁻¹ along the boat. At what speed does the boat move while he is running? At what speed does it move when he stops? If, instead of stopping at the bow the man had dived forward off the bow at the same speed, what would the boat have done then?   (O. & C.*)

43. A uniform circular metal disc with centre O and radius 6 cm has cut from it two smaller discs each of radius 1 cm with centres at points A and B. The distances from O of A and B are each 3 cm and angle AOB is 90°. Find the position of the centre of gravity G of the remaining portion, referred to OA and OB as axes. The two smaller discs are now placed on the larger one so that their centres coincide at O. If G′ is the centre of gravity of this new configuration, show that OG/OG′ = 1.06, correct to three significant figures.

(A.E.B.*)

44. Find the position of the centre of gravity of the frustrum of a cone height 12 cm, if the radii of the two circular faces are 3 cm and 4 cm. (The height of the centre of gravity of a cone, vertical height *h*, is ¼*h* above the base.)

45. A sphere of radius 2 cm has a spherical hole of radius 1 cm cut from it. The centre of the hole is 1 cm from the centre of the sphere. Find the position of the centre of gravity of the remaining portion

46. A ship X at a point O, is moving due east at 10 km h⁻¹. At the same moment another ship Y, at a point 8 km due north of O, is moving due south at 16 km h⁻¹. Find, by drawing, the distance between the ships when they are nearest together.   (C.*)

47. The wind is blowing from the direction 300° (N. 60° W) at a speed of 12 km h⁻¹. Find by drawing or by calculation, the magnitude and direction of the wind as it appears to blow to a man travelling at 8 km h⁻¹ in the direction 090° (due east).
To a man travelling in the direction 180° (due south) the wind appears to be blowing from the direction 270° (due west). Find the speed at which the man is travelling.   (C.*)

48. One ship is sailing due east at 16 km h⁻¹ and another is sailing due north at 12 km h⁻¹. Find graphically the magnitude and direction of the velocity of the second ship relative to the first.   (A.E.B.*)

49. To a man running south-east at 13.5 km h^{-1}, the wind (assumed horizontal) appears to come from the south. When he walks at 6 km h^{-1} in the same direction, the wind appears to come from the south-west. Find graphically the magnitude and direction of the wind.   (A.E.B.*)

50. A ship A is travelling due east at 10 km h^{-1} and at 0900 h is 30 km south-west of another ship B. If B travels at 15 km h^{-1} so as to intercept A in the least possible time, calculate

    (a) the direction in which B must travel,
    (b) the time it takes to the nearest minute, when the interception takes place.

If the speed of each ship is doubled, all other data remaining unchanged, determine the revised answers to (a) and (b) above.
                                                    (A.E.B.*)

51. A river with straight parallel banks is 250 m wide. The current is flowing at 3 km h^{-1} and a swimmer can swim at 5 km h^{-1} in still water. Find, by drawing or by calculation,

    (a) the least time in which a swimmer can reach the opposite bank,
    (b) the time taken to swim from a point on one bank to the nearest point on the opposite bank.   (C.*)

52. A man wishes to row across a river of width 300 m from a point A to a point B on the opposite bank 400 m downstream. The speed of the current is 4 km h^{-1} and the man can row in still water at v km h^{-1}. Find the least value of v for him to be able to reach B. When v has this value, find how long the journey takes.   (C.*)

53. A helicopter whose airspeed is 100 km h^{-1} flies from a point A to a point B due west of A. A steady wind of 30 km h^{-1} is blowing from the north-west. By drawing or calculation, find the resultant velocity of the helicopter.
The helicopter returns from B to A while the same wind is blowing. Find the resultant velocity of the helicopter for the return journey.
                                                    (C.*)

54. An aircraft flies at 400 km h^{-1} in still air. A wind of 50 km h^{-1} is blowing from the south. The pilot wishes to travel from a point X to a point Y north-east of X and then to return. Calculate the direction he must steer

    (a) on his outward journey,
    (b) on his return journey.

If the distance XY is 1000 km, calculate the times of the two journeys.   (A.E.B.*)

55. The liner *Queen Elizabeth II* is steaming due north at 48 km h^{-1} and the liner *United States* is steaming due west at 64 km h^{-1}. At midnight the liner *Queen Elizabeth* is 500 km due west of *United States*. Find

(a) the velocity of the *United States* relative to the *Queen Elizabeth*,

(b) the distance between the liners when they are nearest each other,

(c) the time at which the liners are nearest each other.   (J.M.B.)

56. A cricketer standing on horizontal ground throws a ball with a velocity whose horizontal and vertical components are 16 m s^{-1} and 9.1 m s^{-1} respectively. At the moment when he releases the ball his throwing hand is 1.4 m above the ground. Calculate

(a) the time which elapses before the ball strikes the ground,

(b) the horizontal distance travelled by the ball before it strikes the ground.   (J.M.B.)

57. An aircraft must have a velocity of 50 m s^{-1} at take-off, and it reaches this velocity in 10 s from rest. As its speed increases during its take-off run, its acceleration decreases. Sketch the form of the velocity-time graph, and use it to show that the length of the run cannot be more than a certain number of metres (to be found) but must be at least half this.   (S.M.P.)

58. About 150 years ago a coach was saved from upsetting by the action of an outside passenger, who jumped over a hedge into the adjoining field. Supposing his mass to have been 100 kg, and that just to clear the top of the hedge, 2 m from him horizontally he had to rise 0.1 m in the air, what horizontal velocity would his jump have been capable of imparting to a free mass of 1000 kg?   (O. & C.*)

59. A train of total mass 2×10^5 kg is moving up an incline of 1 in 98 at a uniform speed of 15 m s^{-1}. If the frictional resistances are 32 N per 10^3 kg, find in kW the rate at which the engine is working. If the engine works at the same rate and the frictional resistances are unaltered, find the acceleration of the train when it is travelling on horizontal track at 15 m s^{-1}.   (A.E.B.*)

60. A car of mass 800 kg accelerates on a level road from 10 m s^{-1} to 25 m s^{-1} in 10 s. If the road resistance is 800 N, calculate the tractive force produced by the engine, and the power developed, in kW, when the car is travelling at 25 m s^{-1}.

61. The mass of a cyclist and his machine is 100 kg. When the cyclist is travelling up a hill of 1 in 20 at a steady speed against frictional resistances of 51 N, he is working at 0.4 kW. Find the speed at which he is travelling.

If he works at the same rate along a horizontal road, calculate his acceleration when his speed is 5 m s^{-1}, if frictional resistances are unaltered. What is the maximum speed at which he could travel along this road?

62. Water is discharged from the nozzle of a pump at the rate of 5000 cm^3 every second and the water leaves the nozzle of the pump at a speed of 8 m s^{-1}. Given that the pump is working at 1 kW, calculate the height through which the water is lifted before reaching the nozzle. (C.*)

63. Water is pumped from a depth of 40 m at a rate of 8000 cm^3 a second and ejected through a pipe area of cross-section 40 cm^2. Calculate the work done per second in lifting the water and the kinetic energy acquired per second by the water. (C.*)

64. The maximum power developed by the engine of a car of mass 500 kg is 12 kW. Find the maximum tractive force exerted by the engine when it is travelling at 10 m s^{-1}.
When travelling at 10 m s^{-1} up an incline of 1 in 20 the car can accelerate at 0.6 m s^{-2}. At what rate can it accelerate down the same slope when travelling at 10 m s^{-1}, if the resistances to motion remain the same? (C.*)

65. A man mass 75 kg climbs a height of 200 m in 30 minutes. Calculate his average rate of working against gravity.

66. A body of mass 0.2 kg is attached to one end of a light inextensible string of length 0.25 m, the other end of the string being fixed at a point in a smooth horizontal plane. If the body moves in a circle on the plane and the tension in the string is 5 N, calculate the angular velocity of the body in radians per second. (A.E.B.*)

67. Calculate the additional power that must be exerted by an engine in order to maintain a speed of 10 m s^{-1} when picking up 1000 kg of water from a uniform trough 400 m long. Assume that in being transferred to the engine's tank the water is effectively raised through a vertical height of 2 m. (C.*)

(*This method of taking on more water without stopping was necessary for steam engines, and is still used in parts of the world where steam locomotives are required to travel long distances.*)

68. Unit vectors OP and OP′ make angles α and β respectively with the x-axis OX, where O is the origin. Write down the position vectors of P, P′ and M (the midpoint of PP′) in terms of cartesian unit vectors **i** and **j**.
Express P′ÔM and MÔX in terms of α and β.

Find the magnitude of **OM**. Hence show that

$$\cos \alpha + \cos \beta = 2 \cos \tfrac{1}{2}(\alpha + \beta) \cos \tfrac{1}{2}(\alpha - \beta),$$

and $\sin \alpha + \sin \beta = 2 \sin \tfrac{1}{2}(\alpha + \beta) \cos \tfrac{1}{2}(\alpha - \beta).$

Without using tables show that $\cos 75° + \cos 15° = \tfrac{1}{2}\sqrt{6}.$

(S.M.P.)

69. A body of mass 0.45 kg is attached to a fixed point by a fine string of length 0.8 m and describes a circle uniformly in a horizontal plane containing the fixed point. If the string makes five complete revolutions every three seconds, find the tension in the string.

(A.E.B.*)

70. A body of mass 1.5 kg is attached to a fixed point by a light inextensible string of length 0.5 m and rotates in a horizontal circle on a smooth plane containing the fixed point. If the breaking tension of the string is 40 N, calculate the maximum possible number of revolutions per minute.   (A.E.B.*)

71. A toffee of mass 8 grams is placed on a gramophone turntable at 5 cm from the axis of the turntable. The turntable then rotates at 45 r.p.m. Find the frictional force exerted on the toffee if the toffee remains in position on the turntable.

72. Like parallel forces **F**, **F**, and **2F** act at the vertices of a triangle. Prove that the resultant force passes through a fixed point, called the centre of force, whatever the direction of the parallel forces.

73. Like parallel forces **F**, **F**, **2F** and **2F** act at the corners of a square ABCD respectively. If the length of a side of the square is 6a, find the distances from AB and BC of the centre of these forces.

74. Water collected in a tank at a height of 8 m above ground level is led off by a curved pipe which delivers the water in a horizontal jet at ground level. Find the velocity with which the water leaves the pipe, if one-tenth of the total energy of the water is lost in friction between the water and the pipe.

If the pipe is bent so that it continues to deliver the water at ground level, but the jet is vertically upwards, to what height does the water now rise? Assume the loss of energy is the same as before.

(O. & C.*)

75. A paper-covered book, 10 cm thick when closed, has 200 pages, that is 100 leaves, including the covers. Show that, as each page is turned by the reader, the centre of gravity of the book shifts 1 mm to the left.   (O. & C.)

Miscellaneous Exercise 2

1. An electric train accelerates uniformly from rest to a speed of 20 m s^{-1} which it maintains until the brakes are applied. It is then brought to rest by a uniform retardation equal in magnitude to twice its former acceleration. The total distance covered is 7.8 km and the total time taken is 7 minutes.

Sketch a velocity–time diagram.

Calculate

(a) the time for which the train is travelling at constant speed,

(b) the initial acceleration in m s^{-2}.   (C.)

2. A particle is travelling in a straight line with a velocity, $v \text{ m s}^{-1}$ given by $v = 9 - \frac{1}{4}t^2$ where t is the number of seconds after passing a fixed point O. Calculate

(a) the value of t when the velocity is instantaneously zero,

(b) the distance travelled by the particle during the third second.

    (C.)

3. A particle, moving in a straight line with uniform acceleration, passes a point A with a velocity of 2 m s^{-1}. It reaches a point B after 3 s and a point C after a further 1 s. The distance BC is 16 m. Find the acceleration and the distance AB.

D is the point beyond C where CD is 32 m. Another particles leaves D at the same time as the first leaves A, the second particle travelling towards A with a constant velocity of 5 m s^{-1}. After how many seconds will the particles meet?   (O. & C.)

4. A body starts from rest and moves in a straight line under a constant force. If t is the time in seconds from the instant the body starts, it is found that between $t = 20$ and $t = 30$ the body travels 100 m. Find its constant acceleration and its velocity when $t = 20$. If it continues with the same acceleration, calculate the value of t when it attains a speed of 50 m s^{-1}.

If the original force now ceases to act and a retarding force of 40 N halves its speed in 5 seconds, find the mass of the body.

    (O. & C.)

5. A train which is travelling at a constant speed of 36 m s^{-1} is required to slow down in order to pass through a station. It does so by decelerating uniformly at 0.2 m s^{-2} until its speed is reduced to 12 m s^{-1}, it maintains a speed of 12 m s^{-1} for 1.8 km and then accelerates uniformly at 0.1 m s^{-2} until it reaches its cruising speed of 36 m s^{-1} once more.

Find, in minutes, for how long the speed of the train is less than 36 m s^{-1}, and show that in this time it travels 10.44 km.   (S.M.P.)

6. A continuously varying force acts in a constant direction on a mass of 15 kg. The table gives the value of the total force acting on the mass at certain times.

Force (N)	15	31.5	45	69	75	67.5	34.5
Time (s)	0	0.25	0.5	1.25	2	2.5	3

Calculate the acceleration of the mass at each of the given times. Draw an acceleration–time graph, and hence estimate the change in velocity that the mass undergoes in this 3 second period.

What constant force acting on the mass would produce the same change in velocity in the same time interval?   (M.E.I.)

7. A missile moves so that its velocity at time t after launching is given by $\mathbf{v} = 50\mathbf{i} + (70 - 10t)\mathbf{j}$, where $\mathbf{i}$ and $\mathbf{j}$ are unit vectors horizontally and vertically.
  (a) Calculate its speed when $t = 2$.
  (b) Find the angle between its direction of motion and the horizontal when $t = 2$.
  (c) Calculate its acceleration when $t = 3$.
  (d) Write down its displacement from the launching point at time t.
  (e) State when its height is greatest.
  (f) Find its greatest height above the launching point.
  (g) State, with reason, whether it is a powered missile.   (S.M.P.)

8. A mass of 3 kg has position vector $\mathbf{r} = (3t^2 + 2t)\mathbf{i} - 4t^2\mathbf{j}$, where r is measured in metres and t is the time in seconds.
Find its velocity and acceleration vectors in terms of t.
Show that it is subject to a force of magnitude 30 N, and find the direction of the force.   (S.M.P.)

9. The force acting at time t ($t \leqslant 3$) on a small body, mass 2 units, is $4t\mathbf{i}$. Initially the body is at rest with position vector $2\mathbf{j}$. Find the position vector of the body when $t = 3$.
When $t > 3$, the force acting on the body is $2\mathbf{j}$. Show that, when $t = 6$, the speed of the body is $3\sqrt{10}$, and sketch the path of the body for values of t from 0 to 10.

10. A mass of 20 kg has a constant acceleration vector $\mathbf{a} = -2\mathbf{i} + 10\mathbf{j}$ m s $^{-2}$, where $\mathbf{i}$ and $\mathbf{j}$ are unit vectors in the horizontal and vertically downward directions respectively. Write down, in terms of $\mathbf{i}$ and $\mathbf{j}$, the force in newtons acting on the mass, and hence the impulse in newton-seconds, also in terms of $\mathbf{i}$ and $\mathbf{j}$, given to the mass in a period of 5 seconds. If the initial velocity vector of the mass is given by $\mathbf{u} = 60\mathbf{i} + 70\mathbf{j}$ m s $^{-1}$, calculate the magnitude of the velocity vector after 5 seconds.   (M.E.I.)

11. An object has an initial velocity vector $\mathbf{u} = 24\mathbf{i} + 40\mathbf{j}$, and is subject to a constant acceleration vector $\mathbf{a} = 2\mathbf{i} - 10\mathbf{j}$, where $\mathbf{i}$ and $\mathbf{j}$ are unit vectors in the Ox and Oy directions respectively.
Find the velocity vector $\mathbf{v}$ of the object after
(a) 2 seconds,
(b) t seconds.
In what direction, with respect to the positive Ox direction, is the object initially moving?
At what times will the object be moving in a direction which is at $45°$ on either side of the positive Ox axis? At what time will the object be moving at right angles to its initial direction of motion?
(M.E.I.)

12. A uniform ladder of length 10 m and weight 150 N is leaning against a smooth vertical wall with its foot resting on rough horizontal ground at a distance of 2.8 m from the wall. When a boy of weight 540 N is standing three-quarters of the way up the ladder, its foot is on the point of slipping.
   (a) By taking moments about the foot of the ladder, find the reaction of the wall on the ladder.
   (b) Find, as a fraction, the coefficient of limiting friction between the ladder and the ground.   (O. & C.)

13. A uniform bar AB has length 60 cm and mass 5 kg. The bar, which is freely hinged at A, rests in a horizontal position on a vertical support at C, where AC = 20 cm. Calculate the reaction at C, and deduce the force exerted by the bar on the hinge, stating its magnitude and direction.   (O. & C.)

14. A uniform rod rests horizontally on two fixed supports P and Q which are 60 cm apart. If the rod is of length 2.4 m and mass 12 kg and the reaction of P is twice the reaction at Q, find these reactions and the distances of P and Q from the centre of mass of the rod.
A mass of 4 kg is now placed at the end of the rod nearest Q. How far and in which direction will the rod have to be moved horizontally to make the reactions at P and Q equal?   (O. & C.)

15. Two forces, P and Q, have a resultant of magnitude 12 newtons which is inclined at an angle $30°$ to the force Q. If the magnitude of Q is 8 newtons, find the magnitude of P and the angle between the forces P and Q. (The answers may be found by measurement from a scale diagram or by calculation.)   (O. & C.)

16. (a) Find, by drawing or otherwise, the angle between the lines of action of two forces, of magnitude 10 N and 7 N, which have a resultant of magnitude 6 N.

(b) Three forces acting at a point are as follows: 5 N due East, 8 N due North, 12 N on a bearing of 250° (S 70° W).

Find, by drawing or otherwise, the magnitude and direction of a fourth force, which is in equilibrium with the other three. (O. & C.)

17. A thin sheet of metal is in the form of a rectangle, sides 20 cm and 30 cm, with an isosceles triangle, base angles 45°, drawn on each of two adjacent sides; the base of each triangle is along a side of the rectangle, and both triangles are entirely outside the rectangle. Find the distance of the centre of mass of the sheet of metal from each of the other two sides of the rectangle.

18. A solid body consists of a right circular cone and cylinder of the same radius with the base of the cone in contact with one end of the cylinder, the axes of the cone and of the cylinder being in the same straight line. The density of the material of the cone is double that of the cylinder. If the height of the cylinder is h, find, in terms of h, the height of the cone, so that the centre of gravity of the combined body lies at the centre of the circular base of the cone.   (O. & C.)

19. A square lamina of uniform density is bounded by the x- and y-axes and the lines $x = 8$ and $y = 8$. The rectangle bounded by the lines $x = 6$, $x = 7$, $y = 3$ and $y = 5$ is cut out of the square. Calculate
    (a) the centre of gravity of the remaining part of the lamina,
    (b) the angle between the side corresponding to the x-axis and the vertical when it is suspended from the point corresponding to (3, 0).   (O. & C.)

20. A bowl can be considered as a frustrum of a cone, 20 cm deep, with the circular faces having radii 10 cm and 20 cm. If the bowl is filled with water, find the position of the centre of gravity of the water.

21. A bee which flies through the air at 6 m s^{-1} is aiming to reach a tree 600 m due South of the hive. It happens that the sun is also due South, so the bee wishes to move directly towards the sun; but there is a wind blowing in the direction 052° (that is, approximately from the South-West) at 2.5 m s^{-1}.

Sketch a triangle of velocities, showing the resultant velocity of the bee in the direction due South, and mark the known values on your diagram.

By calculation using the sine rule, show that the bee must aim in a direction 19.2° to the right of the sun.

Calculate how long the bee will take to reach the tree.   (S.M.P.)

22. A submarine at rest fires a torpedo at a ship which, at the moment of firing, is 800 m due north of the submarine. The ship is travelling due east. The submarine commander estimates the speed of the ship to be 25 km h^{-1}; the torpedo travels at 73 km h^{-1}. Calculate

(a) the bearing on which he fires the torpedo and (b) the time that elapses before it reaches his estimated target position.

In fact the ship is travelling at only 20 km h $^{-1}$. Calculate (c) the distance between the torpedo and the ship when the torpedo reaches its estimated target position.   (O.)

23. A boat steaming in a direction N 65° E reaches a point P, which is due south of a lighthouse L, and from which two buoys A and B are observed, both due south of P and 500 metres apart. On proceeding in the same direction for $2\frac{1}{2}$ minutes it reaches a point Q due east of L, from which the buoy A is observed to be S 58.5° W and the buoy B is observed to be S 31.8° W.

Calculate:

(a) the distance AL;

(b) the distance QL;

(c) the speed of the boat.

Give your answers in metres or metres per second.   (O.)

24. A destroyer detects the presence of a vessel at a range of 30 nautical miles on a bearing of 060°. The vessel is steaming on a course of 150° at a speed of 15 knots. If the destroyer steams at 22 knots determine either by drawing or by calculation the course the destroyer must steer so that its velocity relative to the vessel is in a direction 060°.

Hence determine the time taken for the destroyer to intercept the vessel if neither changes course.   (C.)

25. A load of 500 kg is lifted by means of cable through a vertical distance of 40 m. The load is accelerated upwards uniformly from rest at a rate of 0.1 m s $^{-2}$ over the first 20 m and decelerated uniformly to rest over the next 20 m at the same rate.

Calculate

(a) the tensions in the cable during acceleration and deceleration,

(b) the maximum velocity attained by the load.   (C.)

26. A crane can lift a load of 1000 kg by means of a single vertical cable which can withstand a maximum tension of 10^4 newton. What is the greatest possible upward vertical acceleration that can be given to the load?

To speed the process, two such cranes are used, but in this case the two cables lie each at 30° on either side of the vertical. If the tensions in the cables are now not to exceed 7500 newton for safety reasons, what is the maximum upward vertical acceleration that can be given to the load?   (M.E.I.)

27. A lifeboat is launched from rest down a slipway which is inclined at 20° to the horizontal. The resistances to motion (assumed constant) amount to 4×10^4 N. If the mass of the lifeboat is 2×10^4 kg, and the boat reaches the bottom of the slipway with a velocity of 8 m s^{-1}, calculate the time that the boat takes to travel down the slipway.

Immediately after entering the water at the bottom of the slipway, the boat moves horizontally with a velocity of 4 m s^{-1}. Use an accurate diagram to find the magnitude of the impulse given to the boat by the water.   (M.E.I.)

28. A toboggan of mass 15 kg is placed on an icy slope which is inclined to the horizontal at an angle of 18°. It is held in position by applying a horizontal force **R**.

[Friction and other resistances may be neglected.]

(a) Draw a diagram to show the forces acting on the toboggan.

(b) Show by calculation that the magnitude of **R**, correct to 2 significant figures, is 48 N. [Take $g = 9.8$.]

(c) Find the acceleration of the toboggan if the force **R** is now removed.   (S.M.P.)

29. A body of mass 10 kg moves down a slope which is inclined to the horizontal at an angle whose sine is 0.2. The frictional resistance of the slope on the body is 4.6 N. Find the acceleration of the body down the slope.

When the body is first observed, it is moving down the slope with a velocity of 8 m s^{-1}. When it has travelled 12 m further down the slope, it overtakes another body of mass 5 kg moving down the slope at 7 m s^{-1}. The two coalesce and move on together. Find their velocity immediately after the impact.   (O. & C.)

30. A toy truck of mass 2 kg is pulled along the level floor by a string inclined at 30° to the horizontal. When it just moves, explain why the tension in the string is not more than 40 N. Hence prove that, if it does move, the resistance to motion must certainly be less than 34 N.   (S.M.P.)

31. A motor-boat tows a water-skier by means of a horizontal, inextensible rope. When the boat and the skier are moving with a constant velocity the engine of the boat is producing a forward force of 60 N. The mass of the motor-boat and its occupant is 500 kg, and that of the skier 80 kg. What is the total resistance to motion on the boat and the skier?

If these resistances are assumed to remain constant, and are divided between the boat and the skier in the ratio 5:1 respectively, calculate the acceleration of the boat and the skier and the tension in the rope when the forward force provided by the engine of the boat is increased to 1800 N.

If the rope suddenly snaps, by how much will the acceleration of the boat increase?   (M.E.I.)

32. A monkey slides down a light vertical rope. At first it accelerates at 2 m s^{-2} and the tension in the rope is 48 g newtons. It then continues at a uniform speed, and finally, near the bottom of the rope, it retards at 4 m s^{-2}. Find the mass of the monkey and the tensions in the rope during the two latter stages.

The rope is now passed over a smooth peg and a second monkey climbs up the rope from the other end in such a way that both the first monkey and the rope are motionless. If the second monkey climbs with an acceleration of $\frac{1}{3}g$ m s^{-2}, calculate its mass.   (C.)

33. A mass of 5 kg is observed moving instantaneously in a direction 180° with a velocity of 6 m s^{-1}. After a certain interval of time the mass is observed moving instantaneously in a direction 270° with a velocity of 8 m s^{-1}. What is the impulse that has been given to the mass?

If an equal impulse is applied during a further period of time equal to the first interval, find the magnitude of the final velocity of the mass.   (M.E.I.)

34. A body of mass 4 kg slides down a smooth inclined plane from A to B, a distance of 80 m. If the velocity of the body at A is 2 m s^{-1} and A is 10 m above the level of B, find the velocity of the body when it reaches B.

If the lower half of the plane is rough instead of smooth, find what constant frictional force will cause the body to come to rest at B.
(O. & C.)

35. Two balls, A of mass 0.1 kg and B of mass 0.3 kg, are travelling horizontally at a height of h m when they collide head on and they both come instantaneously to rest. If the velocity of A was 9 m s^{-1}, calculate the velocity of B just before impact, and the loss of KE in joules.

They now fall vertically, and when they reach the ground the sum of the kinetic energies of the two balls is 392 J. Calculate the value of h.   (O. & C.)

36. A ball of mass 0.4 kg travelling at 20 m s^{-1} is struck with an impulse of 6 N s. Calculate the speed with which it continues:
(a) if it is struck in the direction opposite to its velocity;
(b) if it is struck in a direction at right angles to its velocity;
(c) if it is struck so as to deflect it as much as possible.   (S.M.P.)

37. A boat of mass 5 tonnes is being driven before the wind at a constant speed of 1 m s^{-1}. It is dragging a sea anchor, and the tension in the cable by which it is attached is 6000 N in the direction horizontally backwards. If the cable suddenly breaks, show that the boat starts to accelerate at 1.2 m s^{-2}.

If the acceleration now decreases at a constant rate and is reduced to zero in 3 seconds, sketch the time–acceleration graph.

Find the speed at which the boat will be travelling at the end of the 3 seconds.   (S.M.P.)

38. (a) A jet of water of cross-sectional area 16 cm^2 moving horizontally at 25 m s^{-1} strikes a vertical wall and immediately falls. Find the force on the wall given that 1 cm^3 of water has a mass of 1 g.

(b) A train of mass 240 000 kg travelling at 0.5 m s^{-1} is brought to rest at a railway terminus by means of the buffers which act with an average force of 60 000 N. How long does this take?   (C.)

39. A shunting engine of mass 20 tonnes pushes two trucks, each of mass 2 tonnes, along a horizontal track. The resistance to the engine is 300 N and to each truck 150 N. If the tractive force produced by the engine is 15 kN, calculate the acceleration of the engine and the trucks. Calculate also the thrust between the buffers of the engine and the first truck.

Subsequently, the two trucks on their own run down a slope whose angle of inclination to the horizontal is α, where $\sin \alpha = 0.1$. What must the resistance now be to each truck if they are to descend at a constant speed?   (M.E.I.)

40. An apparatus for drawing water from a well makes use of the work done by a donkey to which it is connected by a pole. The donkey walks round a circle of 3 m radius once every 20 seconds. If the donkey works at a rate of 0.14 kW find

(a) the average pull exerted by the donkey on the pole,

(b) the mass of water raised through a height of 5 m in 1 hour assuming that 79% of the work done by the donkey is lost.

(C.)

41. A stationary hockey ball of mass 0.18 kg is given a horizontal impulse of 1.5 N s at a free hit. What is the magnitude of the velocity with which it begins to move? Travelling over the ground for 2 seconds, the magnitude of the velocity falls by 10%. What is the force (assumed constant) resisting motion in this period?

At the end of this period the ball rebounds from an opponent's stick in such a way that the magnitude of the velocity is halved, and it moves in a direction inclined at 140° to the original direction of motion. Use an accurate scale drawing to determine the magnitude and direction of the force (assumed constant) which has produced this change of velocity, assuming that contact between the opponent's stick and the ball lasts 0.05 s. (M.E.I.)

42. Two particles are projected simultaneously from the same point with the same speed of 25 m s^{-1}. One particle, P, is projected vertically and the other, Q, at an angle θ with the horizontal such that cos $\theta = 0.6$. Prove that, as long as both particles are in the air, the line PQ is constant in direction. Hence, or otherwise, find the height of P when Q reaches the horizontal through the point of projection. (O. & C.)

43. A golf club imparts to a ball a velocity of 35 m s^{-1} at an angle θ with the horizontal, where tan $\theta = 0.75$. The ball lands first on higher ground at a horizontal distance of 112 m from the point of projection.

Calculate

(a) the time of flight;

(b) the height above the point of projection of the point where the ball lands.

Prove that when the ball hits the ground it is travelling in a direction inclined at approximately 33° to the horizontal. (S.M.P.)

44. The spaceship 'Appleno' of mass 3 with velocity $\begin{pmatrix} 4 \\ 5 \end{pmatrix}$ docks with the spaceship 'Sayes' of mass 2 and velocity $\begin{pmatrix} 3 \\ 4 \end{pmatrix}$.

(a) Show that the common velocity afterwards is $\begin{pmatrix} 3.6 \\ 4.6 \end{pmatrix}$.

(b) Calculate the impulse of 'Appleno' on 'Sayes'.

(c) Calculate whether 'Sayes' has been deflected in the positive or negative sense. (S.M.P.)

45. Relative to an observer O, two aircraft have position vectors $10\mathbf{i} + 7\mathbf{j} + \mathbf{k}$ and $21\mathbf{i} + 17\mathbf{j} + 3\mathbf{k}$, the unit of distance being 10 km. Find the distance between the two aircraft.

46. The position vectors of points A and B relative to an origin O are $3\mathbf{i} + 2\mathbf{j}$ and $4\mathbf{j}$ respectively. Find

(a) the vector OC, where C is the fourth vertex of the parallelogram OACB,

(b) the vector OD, where D is the fourth vertex of the parallelogram OBAD,

(c) the vector OE, where E is the fourth vertex of the parallelogram OABE.

Use vector methods to show that DE is parallel to AB.

47. Three points X, Y, and Z have position vectors relative to an origin O, **x**, **y**, and $k(3\mathbf{x} + \mathbf{y})$ respectively. Sketch the positions of O, X, Y, and Z, and find the value of k if
   (a) YZ is parallel to OX,
   (b) XZ is parallel to OY,
   (c) Z lies on the straight line XY.

48. Forces $\mathbf{i} + 3\mathbf{j}$, $-2\mathbf{i} - \mathbf{j}$, and $\mathbf{i} - 2\mathbf{j}$ acts through the points with position vectors $2\mathbf{i} + 5\mathbf{j}$, $4\mathbf{j}$, and $-\mathbf{i} + \mathbf{j}$ respectively. Show that the vector sum of these forces is zero. Show graphically that these forces do not act through a single point and so cannot be in equilibrium.

49. An UFO is observed to leave a tower from a height of 80 metres and to move so that its displacement t second later is given by $3t\mathbf{i} + 4t\mathbf{j} + 5t^2\mathbf{k}$, where **i**, **j**, **k** are unit vectors West, North and vertically downwards.
   (a) State its velocity after t s.
   (b) Show that initially it is travelling horizontally and calculate its bearing.
   (c) Show that its speed after 2 seconds is $5\sqrt{17}$ m s^{-1}.
   (d) Calculate its distance from its starting point after 2 seconds.
   (e) Find when it hits the ground.
   (f) Calculate its acceleration and hence state, with a brief reason, whether it is powered or not. (S.M.P.)

50. An aircraft takes off from the end of a runway in a southerly direction and climbs at an angle arc tan $(\frac{1}{2})$ to the horizontal at a speed of $225\sqrt{5}$ km h^{-1}. Show that t seconds after take-off the position vector **r** of the aircraft relative to the end of the runway is $(t/16)(2\mathbf{i} + \mathbf{k})$, where **i**, **j**, **k** represent vectors length 1 km, in directions South, East, and vertically upwards.

At time $t = 0$ a second aircraft flying horizontally South-West at $720\sqrt{2}$ km h^{-1} has position vector $-1.2\mathbf{i} + 3.2\mathbf{j} + \mathbf{k}$. Find its position vector at time t. Show there will be a collision unless courses are changed, and find at what time this will occur.

(S.M.P.)

Answers

EXERCISE 1 (page 4)

1. 12. **2.** 360. **3.** 200. **4.** $\pi/2$. **5.** 120π cm s^{-1}. **6.** 45 km h^{-1}.
7. 5. **8.** 1670. **9.** 7.5 km h^{-1}. **10.** 1.27 s. **11.** $33\frac{1}{3}$ km.
12. 5.05×10^3 s. **13.** 7.3×10^{-5}. **14.** $\dfrac{x+y}{T+t}$ km h^{-1}. **15.** $\dfrac{ut+vT}{t+T}$ km h^{-1}.
16. $\dfrac{(x+y)uv}{xv+yu}$ km h^{-1}. **17.** (a) $8\frac{1}{3}$ cm s^{-1}; (b) 5 cm s^{-1}.
18. (a) 4 km h^{-1}; (b) 2.8 km h^{-1}. **19.** $33\frac{1}{3}$ s. **20.** 32 hours.

EXERCISE 2 (page 10)

1. 1.08×10^5. **2.** 10 m s^{-1}. **3.** (a) 20 m s^{-1}; (b) $13\frac{1}{3}$ m s^{-1}. **4.** 6.48.
5. 833. **6.** 16 m s^{-1}. **7.** 2, 4, 6, 8, 10, 12 m s^{-1}. **8.** 4, 6, 8, 10, 12 m s^{-1}.
9. (a) 44.1 m; (b) 24.5 m; (c) 29.4 m s^{-1}. **10.** (a) 20 m; (b) 47 m s^{-1};
(c) 10 s. **11.** 14 m. **12.** 3 m s^{-1}; 19 m s^{-1}. **13.** 20 cm s^{-1}; 10 cm s^{-1}.
14. 955. **15.** 10.5 m s^{-1}. **16.** (a) 20 m s^{-1}; (b) 2 s. **17.** u m s^{-1}.
18. 18 m. **19.** 400 m. **20.** 3 m s^{-2}.

EXERCISE 3 (page 17)

1. 1200. **2.** $\frac{25}{24}$. **3.** $\frac{25}{324}$. **4.** 64 800. **5.** 0.6 m s^{-2}. **6.** 54 m.
7. 35 km h^{-1}. **8.** 18.8 m s^{-1}. **9.** 8 m s^{-2}. **10.** $1\frac{1}{2}$ m s^{-2}.
11. 30 km h^{-1}. **12.** $3\frac{1}{3}$ km. **13.** 2. **14.** 3.5 km. **15.** (a) 1 s; (b) 20 cm s^{-2}.
16. 10 cm s^{-2}; 20 cm s^{-1}. **17.** $\frac{37}{6}$. **18.** 47.5 m. **19.** 72.
20. 6 m s^{-1}; 8 m s^{-1}; 2 m s^{-2}.

EXERCISE 4 (page 24)

1. 14.4 s. **2.** 1.6 km. **3.** 0.04 s. **4.** 28 m s^{-1}. **5.** 1 m s^{-2}. **6.** 75 m.
7. 7200; 4000. **8.** 820 m. **9.** 3000. **10.** $53\frac{1}{8}$. **11.** 100 m. **12.** 4 s.
13. $26\frac{2}{3}$ m. **14.** 39 m. **15.** $\frac{15}{32}$. **16.** 4 m s^{-2}. **17.** 2 m s^{-2}.
18. 20 m s^{-1}; -8 m s^{-2}. **19.** 11 m s^{-1}; 29 m s^{-1}. **20.** 12 m s^{-2}.
22. 5 m s^{-1}. **23.** 24 m s^{-1}. **24.** 144 m. **25.** $\dfrac{60uv(x+y)}{60vx+60uy+tuv}$.

EXERCISE 5 (page 32)

1. 0.45 s. **2.** 19.6 m. **3.** 19.6 m. **4.** 20 m s^{-1} upwards; 78 m s^{-1} downwards. **5.** 1 s. **6.** 1.3 m. **7.** 78.4 m. **8.** 3 s. **9.** 4 s. **10.** 196 m s^{-1}. **11.** 40 m. **12.** Vt metres. **13.** $v^2/2g$, $2v/g$. **14.** 44.1 m. **15.** 3. **16.** 20 m. **17.** 1.5 s. **18.** 1.25 m. **19.** $\frac{1}{4}h$. **20.** $\frac{1}{3}$.

EXERCISE 6 (page 41)

1. 50 m s^{-2}. **2.** 5 m s^{-2}. **3.** 4800 N. **4.** 3.5 m s^{-2}. **5.** 10^5 N. **6.** 1 m s^{-2}. **7.** 15 750 N. **8.** 5250 N. **9.** 0.4 m s^{-2}. **10.** 1.5 m s^{-2}. **11.** 20 000 m s^{-2}. **12.** 13.6 N. **13.** 2.3 kg. **14.** 1.8 m s^{-1}. **15.** 200 N. **16.** 200 N. **17.** 900 N. **18.** 50 000 N. **19.** $(F - R)/m$. **20.** $R + ma$.

EXERCISE 7 (page 47)

1. 4 N. **2.** 0.15 m s^{-2}; 90 N. **3.** 624 N. **4.** 0.04 m s^{-2}; 1500 N. **5.** $\frac{1}{9}g$ m s^{-2}. **6.** 24.9 **7.** 156 N. **8.** 3.5 m s^{-2}. **9.** 2.5 m s^{-2}. **10.** 9.8 m. **11.** 2.8 m s^{-2}. **12.** 1.4 m s^{-2}; 16.8 N. **13.** 2.2 s. **14.** 2.8 m s^{-2}; 12.6 m. **15.** 1.4 m s^{-2}. **16.** 3.5 m s^{-2}. **17.** 10.2 m s^{-2}. **18.** 4.4 N, 5.9 N. **19.** 4.2 m s^{-2}. **20.** 15 000 N. **21.** $\dfrac{Mg}{M + m}$. **22.** $\dfrac{M - m}{M + m}g$. **23.** 19.6; 23.6. **24.** 9.3. **25.** $\dfrac{1}{102}g$.

EXERCISE 8 (page 55)

1. 2.5 m. **2.** 2.4 m. **3.** 10 kg, 1.2 m. **4.** 8 N, 12 N. **5.** 2.8g N; 3.2g N. **6.** 4 kg. **7.** 80 cm from centre. **8.** 18g N, 20g N. **9.** 4 m from one end. **10.** 2$\frac{2}{3}$ m. **11.** 62.1 N; 65.3 N. **12.** 7g N; 7g N. **13.** 5 : 1. **14.** 28.8 cm. **15.** 28g N, 16g N. **16.** 40 kg. **17.** 23g N, 30g N. **18.** 2.5 kg, 28 cm from the centre. **19.** 42.5g N; 37.5g N. **20.** 175 g.

EXERCISE 9 (page 62)

1. 1.2 m. **2.** 1.75 m. **3.** 4 m. **4.** 4 m. **5.** 6 N. **8.** 10 kg. **9.** 196 N. **10.** 16 Nm. **11.** 5g N. **12.** 60 N. **13.** 10g N. **14.** 2$g\sqrt{3}$ N. **15.** 6 Nm. **16.** 25g N, perpendicular to rope. **17.** 5 N, 4 m. **18.** 1.5g N. **19.** 4g N. **20.** 6 m. **21.** $\dfrac{10}{\sqrt{3}}g$ N. **22.** 0.4 N m. **23.** 600 N m. **24.** 10g N. **25.** $\dfrac{4g}{\sqrt{3}}$ N.

EXERCISE 10 (page 72)

4. $4i - j$; $7i$; $-2i - 3j$; $2i - \frac{1}{2}j$. **5.** 1.85, 0.77, 2.8, 3.8. **6.** 1.9; 1.2;
1.2; 4.8. **7.** 1.4; 1.4; 1.4; 2.8. **8.** 7. **10.** $i + \frac{1}{2}j$; $\frac{1}{2}i - \frac{3}{2}j$.
12. 2 : 3. **13.** $\frac{1}{6}$; AX : XB = 5 : 1. **14.** $\frac{1}{3}$; AX : XB = 2 : 1.
15. $2i + j$; $3i$; $(1 + t)i + (2 - t)j$. **16.** $\sqrt{5}$ m s^{-1}; $-250i - 100j$;
300 s; 100 s.

EXERCISE 11 (page 77)

1. $2i - 3j$. **2.** $3i + 2j$. **6.** $3i - 2j$. **7.** $\frac{1}{2}(a + c) - b$.
8. 79°. **9.** 23.7 N. **10.** 1690 N. **11.** 2.8 N. **12.** A couple.
13. $d - a = c - \frac{1}{2}(a + b)$. **14.** $d - a = \lambda(c - \frac{1}{2}(a + b))$.
15. $b + c - a$. **16.** $b + 2(c - a)$. **18.** $\frac{3}{4}a + \frac{1}{4}c$. **20.** $\frac{17}{7}$. **21.** $-\frac{7}{2}$.
22. $\frac{3}{4}a + \frac{1}{4}c$. **23.** $\frac{2}{3}a + \frac{1}{3}c$. **25.** $\frac{2}{3}$. **26.** $(-y, x)$. **27.** $3i - \frac{1}{2}j$.
28. $\frac{1}{7}a + \frac{4}{7}b$. **29.** $\frac{1}{3}a + \frac{2}{3}b$. **30.** $\frac{3}{2}ai + \frac{3}{2}aj$.

EXERCISE 12 (page 88)

1. 6.1 N. **2.** 10 N. **3.** 13 N. **4.** 102°. **5.** 14.3 N. **6.** 8 N. **7.** 051°.
8. 79°. **9.** 23.7 N. **10.** 1690 N. **11.** 2.8 N. **12.** A couple.
13. 3.5 N. **14.** 24.4 N, 041°. **17.** 120°. **18.** 20 N, 66° with AB.
19. 61 N, 25° with BA. **20.** 4 N; 4 N, 30° with AB. **21.** 3.2 N.
22. 1.75 N. **23.** 22.3 N, 21 N. **24.** 3 N. **25.** 11 N. **26.** 6.4 N,
70° 30′. **27.** 4.5 N; 3.6 N at 77° 30′ to horizontal. **28.** 27° 30′;
4.4 N, 6.7 N. **29.** 13.5 N; 47 N at 73° to horizontal. **30.** 0.37
of the way along from A. **31.** $23\frac{1}{2}°$ 9.3 N, 8.7 N. **32.** $W \sin (45° - \alpha)$,
$W \cos (45° - \alpha)$, $45° + \alpha$. **33.** $3.26g$ N, $2.23g$ N; $3.84g$ N, 36.8° with
the vertical. **34.** 12.6 cm, 39 N, 69 N.

REVISION EXERCISES (page 91)

1. 12.3 km. **2.** 12.3 m s^{-1}. **3.** 12.3 m s^{-1}. **4.** 12.3 m s^{-1}.
5. 8.16 m s^{-1}. **6.** 10.04 a.m. **7.** 5.7 m s^{-1}, N 10° E. **8.** 0.78 cm s^{-1},
0.31 cm s^{-2}. **9.** 0.11 cm s^{-2}. **10.** 7 m s^{-1}, N 52° E, 10 m s^{-1}.
11. 12.8 m s^{-1}, $(9 + 6t + 5t^2)^{\frac{1}{2}}$ m s^{-1}. **12.** $(2 + 6t + 5t^2)^{\frac{1}{2}}$ m s^{-1}, 7 s.
13. 10, 1. **14.** 11.15 a.m., 245 km, 390 km. **15.** $(0.9i + 0.03tj)$ m s^{-1};
1.5 m s^{-1}. **16.** 16.3 N. **17.** 16.6 N. **18.** 181 N, 1°.
19. $(-300i - 200j)$ N. **20.** $(r_2 - r_1)/t$. **21.** 1, 2. **22.** 2, -1. **23.** $a = kb$.
24. O, A, B collinear. **25.** $-4i + 3j$; 4, -3; 25. **29.** $r = \lambda(a + b)$.
30. $r = \lambda(yx + xy)$.

EXERCISE 13 (page 100)

1. (a) 18.1 N, 8.45 N; (b) 11.5 N, 9.64 N; (c) −6.13 N, 5.14 N.
2. (a) 6.40 N, 51°; (b) 7.28 N, 74°; (c) 6.71 N, 153°; (d) 7.21 N, 214°;
(e) 5.10 N, 281°.   **3.** (a) 4.84 N, 18°; (b) 6.08 N, 55.3°; (c) 7.53 N,
47.7°; (d) 6.09 N, 84.6°; (e) 4.44 N, 82°; (f) 7.35 N, 94°; (g) 5.21 N,
161.5°; (h) 7.05 N, 349.4°.   **4.** (a) 8.67 N, 24.5°; (b) 6.55 N, 54.4°;
(c) 10.7 N, 41.5°; (d) 3.58 N, 93°.

EXERCISE 14 (page 103)

1. 10.2 N, 79° with AB, 2 m. **2.** $2\sqrt{2}$ N, parallel to CA, through a point
in BA produced, 8 m from A. **3.** $\sqrt{40}$ N, at 252° with AB, 2 m.
4. 4 N, 4 N, 3 N. **5.** 2 N, parallel to BC, $\frac{3}{2}a$. **6.** 2 N, parallel to BC,
$x = 2$. **7.** 0.47 N m. **8.** 0.85 N, 168°. **9.** 6i N. **10.** 12i N, (4i + j).
11. 5.6 N, 44° 20′ east of north. **12.** 4.8 N, 71° with AB. **13.** $2\sqrt{2}$ N
at 45° with BA. **14.** $\sqrt{7}$ N at 41° with BA. **15.** $\sqrt{5}F$ at $\tan^{-1} 2$ with
BA. **16.** A force equal and parallel to AD. **18.** 11.3 N, 87° 30′ with
BA. **20.** 3.5 N. **21.** 2000 N. **22.** A force equal and parallel to CA.
25. 2.4 N m. **26.** 40° 54′. **28.** 90°. **29.** 1 N, parallel to BC,
8 m. **30.** $\sqrt{17}$ N, 75° 58′, 1 m beyond B.

EXERCISE 15 (page 116)

1. 1 N, no motion; 7 N, no motion, 7.84 N, acceleration 0.08 m s^{-2};
7.84 N, acceleration 4.08 m s^{-2}. **2.** 1000 N, forwards. **3.** −147i N,
0.1 m s^{-2}.   **4.** 2.75 m s^{-2}, 14.8 N.   **5.** 98 N, 48 N, 98 N, 80 N,
116 N; 0, 0, 20 N, 24.3 N, 24.3 N, $\mu > 0.3$.   **6.** 128 N, 51.2 N,
1.88 m s^{-2}. **7.** 506 N, 11.8° to direction of motion. **8.** (9i − 10j) N.
9. −686i N, 0.14i m s^{-2}.   **10.** 2 N, 15.7 N.   **11.** 98 N, $\mu > 0.5$.
12. 121 N, 0.38.   **13.** Cube slips: cube stays at rest.   **14.** 255 N,
147 N, 30°, 255 N. **15.** 9.6 N, 19.8 N, 22.8 N. **16.** $\mu = 0.75$, 120 N.
17. At rest: slips, acceleration 1.9 m s^{-2}: at rest. **18.** 31.5 N. **19.** 4.27 N.
20. 2.22 m, 2.8 m, 1.66 m. **21.** 13.4 N. **22.** 1.97 N. **23.** 1.7 N, **24.** 10.6 N.

EXERCISE 16 (page 132)

2. 0.8 m. **3.** 0.9 m. **4.** 0.95 m. **5.** 0.24 m. **6.** 0.65 m. **7.** 0.18 m,
0.08 m.   **8.** 0.156 m, 0.17 m.   **9.** 2.9 cm, 3.2 cm.   **10.** 31°.
11. 2.05 cm, 4.55 cm. **12.** 20.6°. **13.** 3.9 cm, 2.1 cm. **14.** (0, 0).
15. (−1.7, 1.8). **16.** $\frac{1}{2}$i. **17.** (−0.8i − j). **18.** $\dfrac{m + 2M}{m + M}a\mathbf{i} + \dfrac{M}{m + M}b\mathbf{j}$.
19. 1.3 m. **20.** 0.59 cm. **21.** $2\frac{8}{15}$, $1\frac{13}{15}$. **22.** 1.43 cm. **23.** 0.90 cm.
24. −0.042, −0.21, 79°.

EXERCISE 17 (page 141)

1. $(\frac{3}{4}, \frac{3}{10})$. **2.** $(\frac{124}{75}, \frac{254}{105})$. **3.** $(0, -\frac{2}{5})$. **4.** $(1, \frac{2}{5})$. **5.** $(\frac{3}{8}, \frac{3}{8})$. **6.** $(2, \frac{4}{5})$.
7. $(\frac{67}{44}, -\frac{203}{110})$. **8.** $(\frac{1}{2}, \frac{2}{5})$. **9.** $(\frac{8}{5}, 2)$. **10.** $(\pi/2, 0)$. **11.** $(0, 0)$.
12. $(\frac{5}{8}, 0)$. **13.** $(\frac{7}{8}, 0)$. **14.** $(\frac{2}{5}, 0)$. **15.** $(\frac{105}{62}, 0)$. **16.** $(0, \frac{2}{3})$.
17. $(\frac{1}{2}, 0)$. **18.** $(2, 0)$. **19.** $(0, 2)$. **20.** $(\frac{1}{2}\pi, 0)$. **21.** $(\frac{1}{4}\pi, 0)$.
22. $\frac{4}{5}a$ from A. **23.** $(\frac{4}{5}, 0)$. **24.** $(\frac{6}{7}, 0)$. **25.** $10/\pi$ cm from centre, on axis of symmetry.

EXERCISE 18 (page 146)

1. 100 N s, 360 N s, 77 000 N s, 4.2 N s, 10^{-22} N s. **2.** $(9.6i - 24j)$ N s.
3. 12.5 N. **4.** $(12\frac{1}{2}i + 5j)$ N. **5.** $(0.5i + 3j)$ N. **6.** $(8i - 4j)$ N.
7. $(2.4i + 0.6j)$ m s^{-1}. **8.** $(-0.12i + 0.72j)$ N. **9.** 1.1 N s.
10. $+0.42i$ Ns, $-0.42i$ Ns. **11.** 5.88 Ns, 29.4 m s^{-1}. **12.** $(10i - 24.4j)$ m s^{-1}.
13. 28.6 m s^{-1}. **14.** $(15i + 20j)$ m s^{-1}. **15.** $(32i + 20j)$ m s^{-1}; 38 m s^{-1}.
16. 0.02 N, 0.41.

EXERCISE 19 (page 153)

1. 3 m s^{-1}. **2.** 2 m s^{-1}. **3.** 2.5 m s^{-1}, 7.5 m s^{-1}. **4.** 4i m s^{-1}.
5. 0.2i m s^{-1}. **6.** 50 N m^{-2}. **7.** 4.12×10^5 N. **8.** $40\sqrt{2}$ N.
9. 1.6 m s^{-1}, 1 m s^{-1}, $t = 3$, 1.8 m s^{-1}. **10.** $\dfrac{2aT - bT^2}{2m}$.
11. 6.6 m s^{-1}. **12.** 3.3 m s^{-1}, 5.3 m s^{-1}.

EXERCISE 20 (page 166)

1. 900 J, 2.25×10^5 J, 3.125×10^7 J, 5×10^{-4} J, 1.8×10^{-14} J.
2. 8 m s^{-1}. **3.** 0.01 kg. **4.** 48 J. **5.** 2.83 m s^{-1}. **6.** 25.0 J.
7. 3.92×10^4 J. **8.** 660 J. **9.** 1.176×10^6 J. **10.** 2.45×10^5 J. **11.** 7840 J.
12. 0. **13.** 4704 J. **14.** 3920 J. **15.** 7300 J. **16.** 800 J. **17.** 3.2×10^4 J.
18. 1.024×10^5 J. **19.** 3.92×10^4 J, 31.3 m s^{-1}. **20.** 300 N.
21. 1750 N. **22.** 2000 J, 15 m s^{-1}. **23.** 1.5 m. **24.** 6.32 m s^{-1}.
25. 0.784 J. **26.** 45 N. **27.** 90 N. **28.** 100 N. **29.** 98 J, 14 m s^{-1}.
30. 845 J, 13 m s^{-1}. **31.** 17.2 N. **32.** 784 N. **33.** 13 m s^{-1}.
34. 11.8 m s^{-1}. **35.** 129 N. **36.** 1800 J, 750 N. **37.** 8 J. **38.** 0.6 J.
39. 29.4 J. **40.** 72 J. **41.** 6 m s^{-1}, 27 J. **42.** 19.6 kW. **43.** 42 W.
44. 533 kW. **45.** 125 W. **46.** 2220 kW. **47.** 1.6 W. **48.** 180 W.
49. 19 680 W. **50.** 2 kW.

EXERCISE 21 (page 180)

1. 096°, 240 km h^{-1}. **2.** 286°, 245 km h^{-1}. **3.** 073°, 228 km h^{-1}.
4. 086°, 268 km h^{-1}. **5.** 008°, 262 km h^{-1}. **6.** 032°, 218 km h^{-1}.
7. 007°. **8.** 081°, 1 h 37 min; 279°, 1 h 37 min. **9.** 034°, 6°, 1 h 41 min
each way. **10.** 255°, 30 min. **11.** 084°, 276°, 1.8 h, 1.45 h.
12. 1.3 min. **13.** 1.98 min, 2.1 min. **14.** 020°, 2 h 30 min.
15. Not between 038° and 062°.

EXERCISE 22 (page 187)

1. 13 km h^{-1}, from 165°. **2.** 42 km h^{-1}, 191°. **3.** 67 km h^{-1}, 297°.
4. 342 km h^{-1}, 056°. **5.** 10^4 km h^{-1}. **6.** 6.1 mm s^{-1}. **7.** 5.2 km h^{-1}
from N 65° W. **8.** 32°, 10.5 km h^{-1}. **9.** 6 km h^{-1}, S 84° E. **10.** 9.2 min.

EXERCISE 23 (page 192)

1. 78.4 m. **2.** 11 m. **3.** 7.65 m. **4.** 0.06 m. **5.** 2.54 m.

EXERCISE 24 (page 197)

1. $\frac{10}{7}$ s, 10 m, $\frac{20}{7}$ s, 69.3 m. **2.** 6.125 km, 24.5 km. **3.** 60 m. **4.** 0.5 s,
0.025 m. **5.** 8.2 m. **6.** Yes. **7.** 7.06 m, 37.6 m. **8.** 3.97 m, 21 m.
9. 16.4 m, 17.5 m, 5.66 m. **10.** 90 m. **11.** 7.5 m. **13.** 19.6 m.
14. 23° 39′, 66° 21′. **15.** 2 km. **16.** 6.22 m. **17.** 33.6 m s^{-1}.
18. 28 m s^{-1}. **19.** 4.68 m. **20.** 36.9° or 60.3°.

EXERCISE 25 (page 211)

1. 50 N. **2.** 500 N. **3.** 400 N. **4.** 0.82 **5.** 0.59. **6.** 480 N, 0.4.
7. 2.8. **8.** 2.5, −6. **9.** $\frac{7}{12}$. **10.** 80 N. **11.** 5, 8/3. **12.** 20, 55.6 N.
13. 9.2 cm. **14.** 0.8 N. **15.** 8. **16.** 8. **17.** 7100 N. **18.** 743 N.
19. 6350 N. **20.** 37 700 N.

EXERCISE 26 (page 222)

1. 1 N. **2.** 66 N. **3.** 60 N. **4.** 87.5 N. **5.** 4 rad s^{-1}. **6.** 50 rad s^{-1}.
7. 1.44 × 10^4 m s^{-2}. **8.** 0.5 N. **9.** 1.25 N, 0.196 N, 1.26 N.
10. 60°, 58.8 N. **11.** 1.4 rad s^{-1}. **12.** 65°. **13.** 3.16 rad s^{-1}.
14. 3500 N. **15.** 7°. **16.** 0.384 N. **17.** 7 rad s^{-1}. **18.** 2 m s^{-1},
0.19 J, 0.48 N. **19.** 0.82. **20.** 8400 m s^{-1}.

EXERCISE 27 (page 226)

1. 7i, 7 m s^{-1}; −i, 1 m s^{-1}; 3i + 4j, 5 m s^{-1}. **2.** i + 4j, $\sqrt{17}$ m s^{-1};
i, 1 m s^{-1}; 3i + 4j, 5 m s^{-1}. **3.** 12i. **4.** 10i + 10tj. **5.** 5 kg. **6.** 50 kg.
8. −i + 0.5j. **9.** 14i + 8j; 10i + 4j; 18i + 4j. **10.** 4i; 4i; 4i.
11. $\begin{pmatrix} 10t \\ 1 \end{pmatrix}$, $\begin{pmatrix} 10 \\ 0 \end{pmatrix}$. **12.** $\begin{pmatrix} t^3 + 2 \\ 2t^2 \end{pmatrix}$, $\begin{pmatrix} 6t \\ 4 \end{pmatrix}$. **13.** $\begin{pmatrix} 40 \\ 50 - 10t \end{pmatrix}$, $\begin{pmatrix} 5 + 40t \\ 50t - 5t^2 \end{pmatrix}$.

14. 25; 35; 20. **15.** $-6\mathbf{j}$; $-10\mathbf{j}$; $-8\mathbf{i} - 6\mathbf{j}$. **16.** $4\mathbf{i} + 2\mathbf{j}$, $4\mathbf{i} - \mathbf{j}$; $4\mathbf{i} + \mathbf{j}$; $2\mathbf{i} + \mathbf{j}$; **17.** $2\mathbf{i} + 2\mathbf{j}$, $-3\mathbf{i} - \mathbf{j}$; $5\mathbf{i} - \mathbf{j}$. **19.** No. **20.** $6\mathbf{i}$; 123 N.

EXERCISE 28 (page 230)

1. No, yes, no, yes. **2.** 4, -3; 4, 2; -2, 2; 3, 4.
3. $20\mathbf{i}$; $12\mathbf{i} + 16\mathbf{j}$; $-16\mathbf{i} + 12\mathbf{j}$; $10\sqrt{2}\mathbf{i} + 10\sqrt{2}\mathbf{j}$; $\pm(12\mathbf{i} + 16\mathbf{j})$.
4. $5\mathbf{i} + 4\mathbf{j}$; $-5\mathbf{i} - 4\mathbf{j}$; O. **5.** $1 : \sqrt{2}$

EXERCISE 29 (page 233)

1. $\sqrt{50}$; 3; 11; 13, **2.** 3, 11, $\sqrt{170}$; $-2/3$, $2/11$, 0.
3. $1/3$, $2/11$, $1/\sqrt{138}$; $2/3$, $6/11$, $4/\sqrt{138}$. **4.** $\sqrt{(1 - a^2 - b^2)}$.
5. $\mathbf{i} + 3\mathbf{j} + 2\mathbf{k}$. **6.** $13\mathbf{i} + 20\mathbf{j} + 15\mathbf{k}$. **7.** $18\mathbf{i} + 12t\mathbf{k}$.
8. $5\mathbf{i} + 60t\mathbf{j} + 180t^2\mathbf{k}$; 2.2 m s^{-1}. **9.** $6\mathbf{i} + \mathbf{k}$. **10.** $2\mathbf{i} + 20\mathbf{j} - 0.2\mathbf{k}$.
11. $5\mathbf{i} - \mathbf{j} + 5\mathbf{k}$; $7\mathbf{i} - 5\mathbf{j} + 8\mathbf{k}$; $-\mathbf{i} + 7\mathbf{j} - 7\mathbf{k}$; $\mathbf{i} + 11\mathbf{j} + 2\mathbf{k}$. **12.** 3;
$\sqrt{126}$; $\sqrt{89}$; $\sqrt{206}$. **13.** $6\mathbf{j} - 10\mathbf{k}$; $-6\mathbf{j} + 10\mathbf{k}$. **14.** Yes.
15. $3\mathbf{i} + \mathbf{j} + \mathbf{k}$; $3\mathbf{i} - \mathbf{j} + 2\mathbf{k}$; $2\mathbf{i} + \mathbf{j}$.

MISCELLANEOUS EXERCISE 1 (page 235)

1. 0.06 N, 9 s, 6 s. **2.** 10 m s^{-1}, 40 m s^{-1}, $2 < t < 5$, $6\frac{5}{6}$ m, 7 m s^{-2}, 2 or 5,
3. 6.75 mm s^{-2}, 54 mm s^{-1}, 1.08 m. **4.** $(V + na)$ m s^{-1}, $(Vn + \frac{1}{2}an^2)$ m.
$\frac{3}{2}V$ m s^{-1}, 40 V^2/a. **5.** 10 s, 99 m s^{-1}. **6.** $\frac{8}{3}$ m s^{-2}, $33\frac{1}{3}$ m s^{-1}, 50 m s^{-1}.
7. 4.5 min., $\frac{5}{18}$ m s^{-2}. **8.** 60 s. **9.** 0.375 m s^{-2}, 1350 m.
10. $\frac{1}{6}$ m s^{-2}, 8100 m. **11.** Passes through A and C; 10.8 N.
12. 60 N m; 1.88 N, 3° 41'. **13.** 4.47 N, 63.4° with AB 6.5 cm beyond
B. **14.** 6 N, 6 N. **15.** 4.4 N. **16.** 6.24 N, 44°. **17.** 147 N, 153 N,
147 N, 141 N. **18.** 104 N, 0.83 m s^{-2}. **19.** 1055 N, 1225 N, 660 N.
20. 625, 40 s. **21.** 944 N, 784 N, 544 N, 10 s. **22.** $(7\mathbf{i} + 11\mathbf{j})$ N,
$(3.5\mathbf{i} + 5.5\mathbf{j})$ m s^{-2}, 29.3 m. **23.** 1.92 m s^{-1}, 6.4 m. **24.** 3.06 m s^{-1},
15.3 m. **25.** 57.7 N. **27.** 294 N, 104 N, 93 N. **28.** 104 N, 23.6 N,
6.7 m s^{-2}, 0.84 m. **29.** 0.25. **30.** 8.4 m. **31.** 2.5, 0.98 m s^{-2}.
32. 98 N s, 0.6 N. **33.** 1500 N. **34.** 2 m s^{-1}, 1.5×10^5 J. **35.** 6 m s^{-1},
27 J. **36.** $\frac{4}{7}$ s, 5.6 N s. **37.** 3200 N. **38.** 1.6 m s^{-1}. 3.2×10^5 J,
1792 J. **39.** 1 m s^{-1}, 150 J. **40.** 10 m s^{-1}, 35 686 J, 356 860.
41. 0.7 m s^{-1}, 1.47 N s, 0.15 kg. **42.** 1 m s^{-1}, 2 m s^{-1}. continued with
velocity 1 m s^{-1}. **43.** $(-\frac{3}{34}, -\frac{3}{34})$. **44.** 1.36 cm. **45.** $\frac{1}{7}$ cm from centre
of large sphere, on axis of symmetry. **46.** 4.25 km. **47.** 6.46 km h^{-1},
338°; 6 km h^{-1}. **48.** 20 km h^{-1}, 307°. **49.** 9.6 km h^{-1} from 264°.
50. S 17° W, 1.5 h; S 17° W, 45 minutes. **51.** 180 s, 225 s. **52.** 2.4;
9.4 minutes. **53.** 76 km h^{-1} due west, 118 km h^{-1}, due east.

54. N 50.1° E; S 39.9° W. **55.** 80 km h⁻¹, S 53° W; 300 km; 5 a.m.
56. 2 s, 32 m. **57.** 500 m. **58.** 1.4 m s⁻¹. **59.** 396 kW, 0.1 m s⁻².
60. 2000 N, 50 kW. **61.** 4 m s⁻¹, 0.29 m s⁻², 7.8 m s⁻¹. **62.** 17.1 m.
63. 3136 J, 16 J. **64.** 1200 N, 1.58 m s⁻². **65.** 82 W. **66.** 10.
67. 1.74 kW. **69.** 40 N. **70.** 69. **71.** 0.009 N. **73.** $4a$, $3a$.
74. 11.9 m s⁻¹, 7.2 m.

MISCELLANEOUS EXERCISE 2 (page 246)

1. 6 min, 0.5 m s⁻². **2.** 6, $7\frac{5}{12}$ m. **3.** 4 m s⁻², 24 m; 4.5. **4.** 0.4 m s⁻²,
8 m s⁻¹; 125; 8 kg. **5.** 8.5 min. **6.** 11.4 m s⁻¹, 57 N. **7.** $50\sqrt{2}$; 45°;
$-10\mathbf{j}$; $50t\mathbf{i} + (70t - 5t^2)\mathbf{j}$; after 7 s; 245 m; no, acceleration approximately only that due to gravity. **8.** $\mathbf{v} = (6t + 2)\mathbf{i} - 8t\mathbf{j}$; $\mathbf{a} = 6\mathbf{i} - 8\mathbf{j}$;
$\tan^{-1} \frac{4}{3}$ below **i**. **9.** $9\mathbf{i} + 2\mathbf{j}$. **10.** 130 m s⁻¹. **11.** $28\mathbf{i} + 20\mathbf{j}$;
$(24 + 2t)\mathbf{i} + (40 - 10t)\mathbf{j}$; arc tan $\frac{5}{3}$, $\frac{4}{3}$ s, 8 s; $\frac{68}{11}$ s. **12.** 140 N; $\frac{14}{69}$.
13. 73.5 N, 24.5 N downwards. **14.** 78.4 N, 39.2 N, 20 cm, 40 cm;
40 cm in the direction $\overrightarrow{PQ}$. **15.** 6.46 N, 68°. **16.** 144°; 7.4 N, 302°.
17. 17 cm, 13.6 cm. **18.** $h\sqrt{3}$. **19.** $(3\frac{57}{62}, 4)$ arc cot $(\frac{57}{248})$. **20.** 12.1 cm
from smaller face. **21.** 145 s. **22.** 020°, 42 s, 58 m. **23.** 306.4 m,
500 m, 3.68 m s⁻¹. **24.** 103°, 112 minutes. **25.** 4950 N, 4850 N,
2 m s⁻¹. **26.** 0.2 m s⁻², 3.2 m s⁻². **27.** 5.9 s, 8.9 × 10⁴ N s. **28.** 3 m s⁻²
29. 1.5 m s⁻²; 9 m s⁻¹. **31.** 60 N; 3 m s⁻², 250 N; 0.5 m s⁻². **32.** 60 kg,
590 N; 830 N; 45 kg. **33.** 50 N S; 17.1 m s⁻¹. **34.** $10\sqrt{2}$ m s⁻¹;
10 N. **35.** 3 m s⁻¹, 5.4 J; 100. **36.** 5 m s⁻¹; 25 m s⁻¹; $5\sqrt{7}$ m s⁻¹.
37. 2.8 m s⁻¹. **38.** 1000 N, 2 s. **39.** 0.6 m s⁻², 2.7 × 10³ N;
1.96 × 10³ N. **40.** 149 N, 2160 kg. **41.** 0.075 N; 38.3 N, 167°.
42. 20.4 m. **43.** 4 s, 5.66 m. **44.** $\binom{1.2}{1.2}$, positive. **45.** 150 km.

46. $3\mathbf{i} + 6\mathbf{j}$; $3\mathbf{i} - 2\mathbf{j}$; $-3\mathbf{i} + 2\mathbf{j}$. **47.** 1, $\frac{1}{3}$, $\frac{1}{4}$. **49.** $3\mathbf{i} + 4\mathbf{j} + 10t\mathbf{k}$;
$10\sqrt{5}$ m; after 4 s; 10 m s⁻²; no; acceleration approximately that due to gravity. **50.** After 16 s.

INDEX